FROZEN STAR

FROZEN STAR

George Greenstein

George Greenstein

FREUNDLICH BOOKS

New York

Published by Freundlich Books
80 Madison Avenue
New York, New York 10016

Distributed to the trade by The Scribner Book Companies, Inc.
597 Fifth Avenue
New York, New York 10017

Manufactured in the United States of America

Library of Congress Cataloging in Publication Data
Greenstein, George, 1940-
Frozen star.
Includes index.
1. Astronomy—Popular works. 2. Pulsars—Popular
works. 3. Black holes (Astronomy)—Popular works.
4. Cosmology—Popular works. I. Title.
QB44.2.G74 1984 523.1 83-25459
ISBN 0-88191-011-2

Illustrations by Jacqueline Aher

The author is grateful to the following for permission to use original source material as a basis for illustrations:

Figures 10 and 16 from PULSARS, *by R. N. Manchester and J. H. Taylor. W. H. Freeman and Company. Copyright © 1977.*

Figure 17 courtesy of T. H. Hankins and The Astrophysical Journal, *published by the University of Chicago Press;* © 1971 *The American Astronomical Society.*

Figure 54 courtesy of Ethan Schreier and The Astrophysical Journal, *published by the University of Chicago Press;* © 1972 *The American Astronomical Society.*

Figures 55, 56, 57 courtesy of Ethan Schreier and The Astrophysical Journal, *published by the University of Chicago Press;* © 1972 *The American Astronomical Society.*

TO
Barbara

ACKNOWLEDGMENTS

This book could never have been written without the generous assistance of the many scientists who gave willingly of their time in interviews. Some of these interviews appear in the pages which follow; others, equally valuable to me, however, in formulating my ideas, do not. To everyone my deepest thanks: to Brandon Carter, Willy Fowler, Riccardo Giacconi, Stephen Hawking—and thanks to Ian Moss for help with this interview—Richard Huguenin, Richard Manchester, Ethan Schreier, Harvey Tannanbaum and John Wheeler. Jocelyn Bell-Burnell's talk in Chapter 2 was published in the *Annals of the New York Academy of Sciences*, and I am grateful both to her and to the Academy for permission to excerpt these remarks. Chapter 12's account of Chandrasekhar's experiences are culled from a 1977 oral history interview of Chandrasekhar conducted by Spencer Weart of the Niels Bohr Library of the American Institute of Physics, New York, and thanks go to both for permission to use this material. Much of the discussion of Eddington in Chapter 9 was shamelessly lifted from Chandrasekhar's "Verifying the Theory of Relativity," which appeared in the Notes and Records of the Royal Society of London, and I am happy to acknowledge my indebtedness.

Thanks to Amherst College for financial support of this project, and to Robert Novick and the Columbia Astrophysics Laboratory for hospitality where much of the work was completed; thanks to Alan Babb and Mary Catherine Bateson for help with mythological matters, to Ellen Perchonock for patience with seemingly endless revisions, and to Elizabeth West and Walter Pitkin for valuable editorial advice.

CONTENTS

PART THREE

THE CONTEXT

PART ONE

———•———

PULSARS

1

Guest Star

Out for a stroll one starry night, it struck me how rarely we astronomers ever look at the sky.

I was visiting a colleague to put the finishing touches on a piece of research we were doing together. The subject was a small but constant fluctuation in the rotation rates of pulsars. He had spent the last two years gathering the data; I was familiar with a group of theories that had been proposed to account for the jitter. Now we were testing each theory by comparing its predictions with the observations. All completely orthodox—the Scientific Method.

Also part of the scientific method was the incessant personal innuendo that went along with this. We had bits of gossip to exchange about the author of every theory. Ranking high on our list was a certain professor M——. Recently we had returned from an international conference in Germany where, one night, a group of us had visited a local beer hall. M—— had spent the evening drinking Tab.

Nor had he refrained from commenting (often) on the evils of drink. All in all, we were inclined to give his theory a hard time. We had spent the afternoon trying to shoot it down, but much to our irritation it had withstood every test. It was doing better than many of the others.

Now we were off in search of drink and some food. Something strange had just struck our attention, an anomaly revealed by the observations. Data in isolation are meaningless, a collection of numbers. Only in the context of a theory do they assume significance, and not until now, after days of immersion in the theories, had we realized the importance of that anomaly. So far as we could tell not one of them would be able to encompass it. Overhead the stars shone down. We did not look at them. We were looking at our feet as we walked, paused, walked again. We were talking, talking, talking.

At the base of a radio telescope a graduate student is playing a saxophone. The telescope is housed in a white geodesic dome, the surface of which is a mosaic of irregular panels, each different in shape from all the others. The purpose of this "quasi-random structure" is to minimize the dome's interference pattern superposed on the incoming radio waves. It works fine.

It also makes a dandy resonator. The student plays through some favorites: "Confirmation," "The House I Live In." The reed is bad: he tries another. He runs through a quick riff, then settles into "My Favorite Things."

His Ph.D. thesis project is to follow the changes in time of the intensities of quasars—distant explosions which many astronomers believe arise from stars falling into giant black holes. Ideally he would like to observe each quasar in his sample twice a month, but competition for use of the telescope has been fierce and the weather uncooperative. His observing sessions are scheduled months in advance with no possibility of change, and somehow it seems the sky has been perpetually clear except when he has been on the telescope. He has been lucky to get one good observation every few months. Visions fill him of his thesis receding into a void.

But now the weather is crystalline and he has 36 uninterrupted hours of use before him. Already the observation of the first quasar is complete and the second underway. There is time to kill while the data accumulate. He starts on "Lush Life," then wanders up and off on a long improvisation. Music fills the dome—wonderful sounds.

"Nebulosity above the southern horn of [the constellation] Taurus. It contains no star; it is a whitish light, elongated like the flame of a taper, discovered while observing the comet of 1758. Observed by Dr. Bevis in about 1731." So noted Charles Messier, then clerk-assistant to the Astronomer of the Navy (Paris), aged twenty-eight, throughout his life an enthusiastic comet-hunter, later to be dubbed by Louis XV "the ferret of comets." In the course of his observations he came upon many nebulosities—faintly luminous diffuse patches, often somewhat cometlike in appearance. Unlike true comets, however, they did not move across the night sky. Eventually, in order that he not mistake them for the objects of his search, Messier decided to catalog them. In the long course of his life he discovered 16 comets and catalogued 102 spurious nebulosities. Today his catalog remains, and the comets are totally forgotten.

The notation quoted above is the first in Messier's catalog, and the nebula so described has come to be known as Messier 1. Somewhere along the line, because of a fancied resemblance, it picked up a new name: the Crab Nebula. The catalog contains many such irregular nebulae—among other things—and most are what they look like: clouds of gas and dust, lit by a nearby star. The major difference between them and ordinary clouds is size, for they are huge by any earth-bound standards, incomparably larger than the Earth itself.

The Crab Nebula too is a large cloud. However it does not shine by reflected starlight at all.

The Crab is some 60 thousand billion miles across—ten light years—large enough to contain within its body a star (though this star is too faint to have been seen by Messier). It is expanding quite rapidly, at about a thousand miles per second. It expands because it has been blown apart by an explosion—it is the debris of an explosion. Knowing how big the nebula is and how fast it is flying outward, one can calculate how long ago the outburst took place. The answer works out to about 900 years. The exact answer, from completely different evidence, can also be worked out: the explosion was seen when it took place in 1054 A.D.—on the fourth of July.

A photograph of the Crab is shown in Figure I of photo section. It looks like a tumbleweed entangled with cotton. The "tumbleweed" is a complicated skein of glowing orange filaments. They are the remnants of a body exploded into ribbons; they glow because they are still hot. The "cotton," an amorphous blue-white cloud, is more peculiar. It is not really composed of matter in the ordinary sense at all. It arises from electrons traveling at great velocities through a magnetic field that pervades the nebula. The field bends their paths into circles; as they arc, they radiate light.

For years there was a problem with these electrons. No one knew where they came from. It was not so difficult to imagine the 1054 explosion sending forth a spray of rapidly moving particles, but they would not have kept their high velocities for long. As they radiated energy they would have slowed, long ago having come essentially to rest and ceasing to emit light. Something else must be producing them and injecting them into the nebula—something going on right now.

There was another problem. Dating the time of the explosion from the rate of expansion of the nebula did not exactly give the right answer. The discrepancy could only be resolved by supposing it had not been expanding at a constant rate. It must have been accelerating. An ordinary cloud of debris slows down as it flies apart; the Crab was speeding up. Again, something must be going on within it—now.

As for the star within the nebula, it was not a star at all. It was a pulsar. But no one realized this till recently.

To those of us who work on pulsars and black holes they are the stuff of everyday life—but only in the sense I have described above. The scientist immersed in research is more bound up by the methods he employs than by the object of his study. The observational astronomer works at his telescope, the theoretician at his mathematics, and both spend an inordinate amount of energy on arguments with colleagues. Somehow in all this the pulsar and the black hole seem to get lost.

It takes a positive effort for me to detach myself from the business of science and visualize the strange beast I have been living with for so many years. But give it a try.

In my imagination I am floating weightlessly in space. Alongside me is an enormous orange wall stretching endlessly into the distance. It is part of a filament in the outskirts of the Crab Nebula. On all sides I am enveloped in milky-white fog. It is emission from the electrons. Stars are visible in every direction, for the nebula is transparent, so diffuse it is very nearly a vacuum. Nevertheless it is filled with intense levels of radioactivity. Radiation consists in part of high-velocity particles, in this case the same that produce the nebular fog, and without massive shielding I would accumulate a lethal dose within seconds. It produces a noticeable heating effect.

Light years away, the pulsar at the nebula's core is a stroboscope. Its light comes in bursts, a brilliant flash and then a fainter one: sixty pulses per second in a steady rhythm. From the outskirts of the nebula the object responsible for the bursts—the "lighthouse" emitting the flashes—is too small to be seen. The largest telescope we have would not be capable of spotting it from there.

I move in toward the heart of the Crab. The level of radioactivity rises. It is concentrated toward the pulsar—it comes from the pulsar. The radiation exerts a noticeable pressure, and it is this force that accelerates the nebula's expansion. As for the pulsar, a moving picture shot from here would reveal something of its nature to me. Played in slow motion the movie would show two beams of light, one brighter than the other, pointing in nearly opposite directions, spinning wildly at thirty rotations per second.

I move still closer. I move to within 93 million miles of the pulsar, the distance of the Earth from the Sun. The pulsations are overwhelming. On average the illumination is brighter than sunlight: concentrated into bursts, the level of each is blinding. A storm of radiation—electrons and protons—pours from the pulsar outward into the nebula. No shielding imaginable would suffice to protect against it. It would be enough to shred a planet. It may even be that long ago a system of worlds did swing about the pulsar in steady orbits, but if so they would have been on fire, violently boiling under the impact of the terrible radiation. From each a vast plume of vaporized rock would have streamed outward. The planets would have looked like comets. None could have survived till now.

The pulsar is as massive as the Sun, though very much smaller— so much smaller that even from so close a telescope would still be unable to find it. As a result of its mass it exerts a force of gravitational attraction on me. I weigh something. Bit by bit I fall inward— and as I do the force of gravity increases. I fall at an ever-increasing rate. At a distance of one million miles from the pulsar I am plummeting at more than 200 miles per second. At a tenth this distance gravity is so strong it pulls a pea with a force of one pound, and the rate of fall has built to 800 miles a second. A steady flow of X rays is pouring outward. It grows stronger as I plummet, as does the radiation level and the brilliance of the light. Ten thousand miles up, the attraction on the pea is a hundred pounds and my speed is more than 2,000 miles per second. Within the time it takes to read this sentence I have dropped to a thousand miles' altitude. There is a powerful magnetic field. It is fluctuating violently—it is *rotating*, precisely in step with the pulsar beacons. I am enveloped in cosmic fire: superheated plasma, intense electric currents. Massive lightning strokes flare against me. Somewhere within this region the spinning searchlight beacons originate. The force on the pea is five tons.

There are now one thousand miles left to be covered to the pulsar, which is still invisible, and I am going to travel this distance, whether I like it or not, in an eighth of a second. Everything happens at once. I am violently accelerated to a sizable fraction of the velocity of light in my fall. Things ahead are blue, things behind red. The rotating magnetic field grows to such unthinkable intensities that atoms are deformed. Unfamiliar effects of gravitation come into play: my body is stretched, geometry is distorted, and the paths of light rays bent. And then the object of this journey and the cause of everything I have experienced bursts into view. There is perhaps a thousandth of a second in which to view it before I rush past. So enormous is my speed by this point that no engine imaginable could alter in the slightest my headlong rush.

It is a magnet, the strongest known to exist, quite possibly the strongest magnet that does exist anywhere in the universe. Ten miles in diameter, it is spherical with a surface so smooth it would take a microscope to see any irregularities. It is spinning about its axis thirty times per second, and is so hot that it glows not red like hot metal, not white like the stars, but in X rays. It has an atmosphere a few inches thick that is violently streaming away into space.

Now I am past it. It was a neutron star.

In the above account there are two things that furnish the key to the nature of neutron stars: the star's mass was that of the Sun, but its diameter a mere ten miles. It must have been very dense. Divide the mass by the volume to find the density. One gets 100,000,000,000,000 times that of water.

That is an interesting number. It is the density of an atomic nucleus.

In ordinary situations matter is composed of atoms—all matter: in a rock, in a living organism, water in the ocean. An atom, in turn, contains a small, dense nucleus with roughly equal numbers of neutrons and protons, and about the nucleus a cloud of electrons. These electrons lie quite far from the nucleus, so that an atom forms a very open, airy structure. If one were to imagine an atomic nucleus the size of a golf ball, the electrons would be found several miles away. Even the densest matter on Earth, say a block of lead, is almost entirely empty: the atoms are tightly packed, but they themselves are nearly vacant spheres.

The imaginary journey recounted above ended with a glimpse of one of the few things in the universe that is *not* mostly empty. Long ago, an intense compressional force had been applied to a star. The atoms were crushed, the nuclei forced together. Under the compression the electrons reacted with the protons within the nuclei to form neutrons. There resulted a tiny, ultradense sphere of neutrons: no atoms, no more empty spaces. A neutron star.

In order to create such "neutronic" matter, all one need do is sufficiently crush ordinary matter. A spoon could be transformed into the neutronic state if only we had vises strong enough. Once created, such matter would be very strange stuff indeed. A chunk the size of a sugar cube would weigh 100 million tons. Placed on a table, the chunk's weight would punch it through the table. It would fall to the floor and punch through that. It would bore through the solid ground and dig a hole down to the center of the Earth, there to overshoot, fly out to the other side, come to rest, and fall inward again; penduluming back and forth within the body of the Earth.

Neutronic matter could only survive on Earth if wrapped in a high-pressure jacket, for it possesses an enormous internal pressure. In a neutron star it is gravity that contains the pressure. On Earth, unless contained by some outside agency, the sugar cube would explode, with a force equivalent to 100,000,000,000 megatons of TNT.

Perhaps it is just as well we cannot produce such stuff. The compressional force required to make it is greater than anything that can be achieved on Earth. But nature might be able to create what we cannot. Gravity could do it. The Sun at present is subject to an intense compressional force arising from the gravitational attraction of each of its parts on every other. This fails to compress it into a neutronic state simply because it is so hot: its high temperature keeps the Sun expanded into a large, diffuse globe. In the Sun this balance is stable, but in other stars it can be precarious. If it were to become upset in some star, what would happen? Would the star collapse into the neutronic state—to become a neutron star?

When a large body shrinks it releases energy, and the more it shrinks the more energy is produced. It is easy to compute how much would be made available if a typical star, with a diameter of about a million miles, were to collapse to become a neutron star, with a diameter of ten miles. The answer is unnerving: more energy would be released in a matter of seconds than the star has radiated throughout all its billions of years of previous history. It would be released explosively. The outer regions of the star would be blasted away while the interior fell into the neutronic state. This process, the destruction of an old star and the simultaneous creation of a new, would be one of the most violent and catastrophic events imaginable.

From the history of the Sung Dynasty by T'o-T'o: "In the first year of the period *chih-ho*, the fifth moon, the day *chi-ch'ou* [a guest star] appeared approximately several inches southeast of T'ien-kuan. After more than a year it gradually became invisible."

The period *chih-ho* corresponds to 1054 A.D.; the fifth moon, the day *chi-ch'ou* is July 4. T'ien-kuan is a region of the sky somewhere near a star we know nowadays as T Tauri—so named because, like the Crab Nebula, it lies in the constellation of Taurus. But what is a guest star?

A further record states: "It was visible by day, like Venus . . . Altogether it was visible [in daylight] for 23 days." The star apparently dimmed steadily: by April 17, 1056, it was no longer even visible at night. So far as we know, it has never reappeared.

Yang Wei-te, presumably chief astrologer of the Sung court, noted that the guest star had "an iridescent yellow color. Respectfully, according to the dispositions of the emperors [the imperial color was yellow], I have prognosticated, and the result said: the guest star does not infringe upon Aldebaran; this shows that a plentiful one is Lord, and that the country has a Great Worthy: I request that this [prognostication] be given to the Bureau of Historiography to be preserved." The event had political overtones as well.

In fact, guest stars are not so uncommon as might be thought. What was uncommon about this one was its brightness. With the advent of astronomical telescopes it has become clear that from time to time certain stars suddenly and unexpectedly flare up, as if they had exploded, only slowly dimming to their original states. If such an exploding star had been originally too dim to be noticed, it would seem to be created upon its sudden brightening. Hence the modern term, *nova*, Latin for "new." The Chinese term *guest star*, was equally graphic: the star briefly visits, then leaves.

A nova does not involve the complete, or even partial, destruction of a star. It is a transient flare-up after which the star settles down into a perfectly ordinary state. But once in a long while—a very long while—a nova of a wholly different aspect appears. These supernovae, as they are called, are very rare; the last to take place even remotely close to us was discovered by Kepler in 1604. To observe such events modern astronomers are forced to examine astronomical photographs of distant galaxies. When one is found it is truly extraordinary. The star—previously invisible, for no individual stars can be seen at such great distances—brightens until it is perhaps one hundred billion times brighter than usual. Slowly it dims, often remaining visible through the finest telescopes for several years before finally vanishing from sight. At its brightest a supernova puts out more light than an entire galaxy. Such an outburst is easily sufficient to disrupt a star. If the Sun were to supernova, all life on Earth would be wiped out in an instant; the very planet would be badly damaged, possibly shattered.

When a supernova is discovered, the news of its appearance is immediately telegraphed to astronomers throughout the world. Since any given galaxy will be photographed only occasionally, the chances are small that a supernova will be found in the initial stages of explosion. Usually it is caught a matter of weeks after the outburst. Many outbursts seem to have similar properties, so similar that they are grouped together as a type. These are characterized by a rapid brightening of the star, a maximum brightness perhaps one hundred billion times that of the Sun, a fairly rapid dimming for about a month, and finally a slower decline into obscurity that seems to persist indefinitely. The spectrum of the light emitted by this type—the detailed analysis of its color—constitutes one of the longest-standing mysteries in astronomy: it has never been interpreted successfully.

Because supernovae are so rare, the only ones seen are very far away—like lightning, which usually strikes somewhere else. Because they are far away, very little information has been gathered about them, and as things now stand they are mysterious things. The supernova explosion of a nearby star would be an unparalleled opportunity. Optical telescopes, radio and X-ray telescopes, cosmic ray detectors, neutrino detectors, gravitational wave detectors . . . the whole arsenal would be pressed into emergency service. In a year, in the very first few days, more might be learned than in all the decades since their discovery. It is hard to estimate the chances of this happening soon, though. Supernovae seem to occur a few times per century in any given galaxy. Since the last nearby one in ours was in 1604, we are long overdue for another. It—or they—may have already come and gone, however, for much of our vicinity is obscured from view by dust clouds so opaque that not even a supernova could be seen through them. All in all, astronomers are not holding their breaths.

It was in 1934 that the astronomers Walter Baade and Fritz Zwicky first realized the true nature of supernova explosions, compared with which the steady shining of billions of stars is utterly insignificant. Just two years before, the Russian physicist Lev Landau had proposed that some kind of "neutronic" matter might exist in the cores of ordinary stars. We now know that Landau's original arguments were fallacious, and had almost no connection with the present concept of a neutron star. Nevertheless, they served to introduce the idea. Baade and Zwicky seized upon it. They wrote in a technical journal: "With all reserve we advance the view that supernovae represent the transitions from ordinary stars into neutron stars, which in their final stages consist of extremely closely packed neutrons." Coming as it did only two years after the discovery of the neutron, the suggestion was an audacious one. As if worried they had gone too far out on a limb, Baade and Zwicky added elsewhere: "We are fully aware that our suggestion carries with it grave implications regarding the ordinary views about the constitution of stars and therefore will require further careful studies."

Nobody paid any attention.

The history of science is strewn with prescient suggestions ignored for decades. This was one such. The old image comes to mind of the scientist as embattled, struggling to bring his ideas to light against a general indifference. It is a perfectly valid one. On the other hand, for each right idea there are a hundred wrong ones: the scientist who spent his career chasing down every highly speculative proposal that came his way would not get very far. And at any rate, Baade, Zwicky, and Landau hardly fit the stereotype, for they themselves had no great faith in their suggestion. Baade never went back to it again; Landau gave it one brief mention in a textbook. To them it was just another notion. Only Zwicky continued to push for the idea occasionally—but he was generally regarded as something of a Wild Man anyway. It is possible to count on the fingers of one hand the technical papers published on neutron stars in the generation following the first speculation.

If the study of neutron stars languished in these years, that of diffuse nebulae—the Crab, for instance—did not. As far back as 1921 it had been included in a list of nebulae that were near—possibly at—the locations of old "novae." In 1942, with the publication in English of the ancient Chinese records quoted above, it became clear that the 1054 guest star had been a nearby supernova. Furthermore, its location coincided very nicely with that of the Crab.

The Crab Nebula was therefore what a supernova looks like 900 years later: it was the debris of an exploded star. There was a star in the Crab. Was it a neutron star? It was obviously unusual, for its spectrum defied analysis and matched that of no other star. But such peculiar stars were not so rare as all that, and it attracted only mild attention.

A few papers were published attempting to interpret it. It is amusing and infuriating to read these old articles: it was this kind of star, it was that kind of star, it was a neutron star. . . . We know now, of course, that it is none of these: it is a pulsar, a rotating searchlight beacon, and this pulsar is somehow a manifestation of a spinning neutron star that to this day no one has been able to detect.

If only the pulsar had been bursting less rapidly! It can be seen in Figure I, its myriad pulses summed up into a point—it looks like a star. It can be seen through a telescope, and as a moving picture, a series of still photos, appears to flow continuously, so it seems to shine steadily. No highly technical equipment is needed to detect the pulses. They could have been found in 1934.

But nobody did it. The "star" remained a star.

2

The Discovery of Pulsars

By the late 1960's, the idea of a neutron star had been in the air for more than three decades. They were eventually discovered by accident.

The accident occurred in Cambridge, England, in 1967, and it happened when the British astronomer Antony Hewish built a new kind of radio telescope. Hewish was not looking for neutron stars at all. He had no particular interest in them. He was interested in twinkling—or, in more technical terms, scintillation. Hewish wanted to study the scintillation of the radio signals from the quasars. To do so he was forced to build a new kind of radio telescope: one sensitive to the minute, rapid fluctuations the scintillation produced. This telescope, in fact, was the very first capable of detecting such rapid fluctuations in the intensity of a cosmic radio source. Purely by accident the design was ideally suited to the discovery of pulsars. For this discovery, as well as for a lifetime of distinguished work in radio astronomy, Hewish received the Nobel Prize in physics in 1974.

It was not Hewish himself who actually found the first traces of the pulsars. This honor goes to a graduate student, Jocelyn Bell. It was she who first found, in late September of 1967, an anomaly buried within the veritable maze of data produced by the new telescope. Initially she did not know what to make of this anomaly. It hardly seemed real. She called it "scruff."

Bell attempted to pin down its nature by more carefully observing the radio source. The source refused to cooperate. It went away. For two full months she went out to the telescope each day, looking for the scruff, and for two months the radio source remained invisible. Many other scientists—and certainly many other students—would have given up. Bell continued. Finally, in late November of 1967, the radio source reappeared, and she immediately realized that what she had been calling scruff was actually a series of regular pulsations. She had discovered the first pulsar.

Years later, in an after-dinner speech at a scientific meeting, Jocelyn Bell—by then Jocelyn Bell Burnell—described her experiences during that wonderful time.

"I joined [Hewish] as a Ph.D. student when construction of his telescope was about to start," she said. "The telescope covered an area of four and a half acres—an area that would accommodate 57 tennis courts. In this area we put up over a thousand posts, and strung more than 2,000 dipoles between them. The whole was connected up by 120 miles of wire and cable. We did the work ourselves—about five of us—with the help of several very keen vacation students who cheerfully sledge-hammered all one summer. It took two years to build and cost about 15,000 pounds, which was cheap even then. We started operating it in July, 1967, although it was several months more before the construction was completely finished.

"I had sole responsibility for operating the telescope and analyzing the data, with supervision from Tony Hewish. We operated it with four beams simultaneously, and scanned all the sky between declinations +50 and −10 once every four days. The output appeared on four 3-track pen recorders, and between them they produced 96 feet of chart paper every day. The charts were analyzed by hand—by me. We decided initially not to computerize the output because until we were familiar with the behavior of our telescope and receivers, we thought it better to inspect the data visually, and because a human can recognize signals of different character whereas it is difficult to program a computer to do so.

"After the first few hundred feet of chart analysis I could recognize the scintillating sources [i.e., the quasars], and I could recognize interference. Six or eight weeks after starting the survey, I became aware that on occasions there was a bit of 'scruff' on the records, which did not look exactly like a scintillating source, and yet did not look exactly like man-made interference either. Furthermore, I realized that this scruff had been seen before on the same part of the records—from the same patch of sky."

Talking it over, Hewish and Bell came to the conclusion that the source of these anomalous signals deserved close attention. They decided to obtain high resolution recordings with a special extra-rapid chart recorder. The radio telescope was only capable of observing sources as they passed overhead in the daily rotation of the sky: she was forced to arrange her schedule around this rotation in order to be present at the telescope whenever the source was detectable. "Towards the end of October, when we had finished doing some special test on [the quasar] 3C273, and when we had at last our full complement of receivers and recorders, I started going out to the observatory each day to make the fast recordings. They were useless. For weeks I recorded nothing but receiver noise. The 'source' had apparently gone.

"Then one day I skipped the observations to go to a lecture, and the next day on my normal recording I saw the scruff had been there. A few days after that, at the end of November '67, I got it on the fast recording. As the chart flowed under the pen, I could see that the signal was a series of pulses, and my suspicion that they were equally spaced was confirmed as soon as I got the chart off the recorder. They were one and one-third seconds apart. I contacted Tony Hewish who was teaching in an undergraduate laboratory in Cambridge, and his first reaction was that they must be man-made. This was a very sensible response in the circumstances, but due to a truly remarkable depth of ignorance I did not see why they could not be from a star. However, he was interested enough to come out to the observatory at transit-time the next day, and fortunately (because pulsars rarely perform to order) the pulses appeared again. "This is where our problems really started."

It is one thing to discover a series of radio bursts. It is another matter entirely to understand what they are. As things then stood only one thing was clear: that Bell had identified a highly unusual source of radio emission. In the long run it turned out that her discovery marked a minor revolution in astronomy. It was at once the culmination of a thirty-three-year-long story and the beginning of another. But no one knew this at the time.

No one knew what it was that emitted these signals. Was it a star? A galaxy? Or was it something entirely different—an object the likes of which no one had even suspected till then? If it was a star or a galaxy, why were its emissions so unlike those of other stars and galaxies? Was there only one of them or were they common?

It was a long way to the final resolution of these questions.

Initially, it was Hewish, Bell, and their colleagues who sought to understand the pulsations' significance. But they did not get very far. Eventually, without having reached a resolution, they published the announcement of their discovery, and with this announcement the problem passed from their hands to those of the world-wide community of physicists and astronomers as a whole. After this point what had initially been an effort confined to a few became an effort conducted all over the world. It was conducted in scientific papers in the journals; conducted by personal letter and by phone; conducted endlessly in universities as far apart as Moscow, Sydney, London, and New York; in department seminars and over lunch.

How was it done? There is no simple answer to this question short of recounting what actually happened. It is a long story, encompassing the efforts of scientists the world over, and looking back in retrospect it resembles nothing so much as the solution of a gigantic crossword puzzle. It was a matter of piecing together clues.

These clues were contained within the radio pulsations themselves. The pulses were not entirely anonymous. Rather their very structure betrayed the nature of their source. It was the business of the scientists to find more such clues and to weave them together into a coherent picture. It took a little more than a year.

At the outset it seemed to Hewish and Bell that the bursts of radio emission were not astronomical in origin at all. Most likely they merely emanated from some man-made source. They appeared too regular to be natural. The spark plugs of a car, for example, radiate just such regular bursting radio signals with every firing. The signals might also have been produced by an electric clock, or by any of a myriad other possibilities. Arguing against this, however, was the fact that the pulsations appeared in the output of their radio telescope only at a certain time of day; the obvious interpretation was that they were coming from a celestial source and that this source passed over the telescope, and hence was observed, at just this time. It all held together.

But this was not the only possible interpretation. Perhaps the bursts were artificial in nature after all, but only turned on at certain selected times of day. Perhaps it was a radio signal meant to actuate a noon factory whistle, or a ham radio operator of exceedingly regular habits. How to distinguish these two interpretations? How to distinguish the time kept by the sky from the time kept down below?

Actually these two time systems are different. This may not be immediately obvious, for the nighttime sky appears to rotate overhead quite steadily. But although the sky's rotation is regular, it does not occur at the same *rate* as that of the hands of clocks. The easiest way to see this is to fix attention on some particular region of the sky—some obvious constellation, or a bright star—and ask whether it passes overhead on succeeding nights at exactly the same time as determined by clocks.

It does not. Each night this landmark passes overhead a little earlier. Sirius, for instance, the brightest star, passes overhead at midnight in late December, but by spring it does so just after sunset. All through the summer Sirius is up in broad daylight, and hence is invisible, and not until the fall does it become visible again early in the morning. Earthly clocks keep human time: the sky keeps *sidereal* time.

To distinguish the two it was only necessary for Bell to observe the scruff over a moderately long period of time. Did it reappear on the chart recordings at exactly the same time each day? Checking the yards of data that had accumulated over the months, Bell realized that it did not. It was keeping sidereal time.

At about the same time John Pilkington, a third member of Hewish's group, succeeded in measuring the *distance* to the pulsar. He did this by observing the object at a new frequency. Radio telescopes, like ordinary radio receivers, operate at a particular frequency of the electromagnetic spectrum, and up to that point all the observations of the pulsar had been made at one frequency. Pilkington tried another. One changes the frequency at which ordinary radios work simply by adjusting the tuning knob. For a radio telescope the procedure is not easy, but it is straightforward enough for all that. Pilkington decreased the operating frequency of the telescope. But upon doing so he found that the bursts at lower frequencies arrived slightly *after* the higher-frequency bursts.

It immediately occurred to Pilkington that this phenomenon very likely did not originate within the pulsar itself. He felt it more reasonable to assume that the source emitted all frequencies at once. He felt this because he was aware of a peculiarity of the propagation of radio signals through interstellar space, and he realized that this peculiarity could give rise to the very phenomenon he had observed. Radio waves propagate at the speed of light—but only in a vacuum, and interstellar space is filled with a faint residual trace of gas. This gas slows the waves. In fact—and this was Pilkington's point—it slows them selectively. It acts upon the signals according to their frequency. The lower the frequency of the wave, the more it is retarded, and the later it arrives at the Earth.

With this interpretation in hand, it was easy for Pilkington to determine the distance to the pulsar. He was faced with a situation very much like that of a race between two runners, one slightly faster than the other. If the race is a short one—a fifty-yard dash—the faster runner will arrive just barely ahead of the slower; a fraction of a second at most. In a long race, on the other hand—ten miles, for example—the winner might cross the finish line a good five minutes ahead of the runner-up. The longer the distance traveled, the greater will be the gap in time between the arrival of the faster and the slower runners.

All Pilkington had to do was determine how much *sooner* the higher frequency signal arrived than the lower, and then apply the very same logic. He did this. He found that the pulsar was one thousand light years away.

As time passed Hewish and his colleagues found their thoughts turning in a somewhat ominous direction. The pulsations they had found were unnervingly steady. They were *too* steady. Each pulse of radio waves arrived one and one-third seconds after the preceding one. More precisely, they arrived every 1.3373011 seconds, and they maintained this inflexible progression, this perfect regularity, with a constancy seldom found in the natural world. If it was a clock they had discovered, it was a very well-made clock indeed.

Perhaps it was too well-made. Perhaps it was too well-made to be natural.

For after all, when does the natural world ever present us with a phenomenon of perfect regularity? More often than not, nature is chaotic and erratic. Standing alone in the forest one hears a multitude of sounds—but hardly ever a steady, an utterly uniform series of clicks. Is it a woodpecker? More likely it is a hidden clock. By and large, patterns of great order are not the product of natural processes at all. They are the marks of intelligence.

Had Jocelyn Bell discovered an extraterrestrial civilization?

It was a daunting prospect. It was one thing to discover a new and unusual source of radio waves. It was another matter entirely to discover an alien intelligence. Strange as it may sound, there is an element of fear in all scientific discovery. It is compounded of many ingredients. There is the fear of going too far out on a limb, of laying claim to an important discovery when the evidence is not yet complete. There is the fear that one has made some subtle but terrible mistake. And there is the fear of success. There is always the fear—hidden, unstated, lurking in the background—of stumbling upon some discovery so important, so earth-shaking, that it changes the world forever. Hewish, Bell, and their coworkers knew that if they were to claim to have discovered evidence of an extraterrestrial civilization in error, they would be the laughingstock of the scientific community. And if they were right, they would be the authors of a discovery that without exaggeration could be termed one of the most momentous in the history of science.

The signals looked for all the world like pulses of light from a lighthouse. Had Bell discovered some sort of navigational aid, warning interstellar travelers of a celestial menace to shipping? They also looked like pulses of emission from the spark plugs of a car. Had she picked up on the emissions from an unseen spaceship passing in the night? Was she eavesdropping on a celestial dialogue? Or were these signals meant for *us*? Were they intentionally beamed at the Earth by some society anxious to establish contact in an effort to alert us to their presence?

Standing against all this was one important fact. The signals were at the wrong frequency to be artificial. They were at the frequency at which *other* sources of cosmic radio signals emitted most strongly. The quasars, the radio galaxies, supernova remnants, our very galaxy itself—all these natural emitters competed with the pulsars. Other, more quiet bands were available, but the pulsars had not selected them. It would hardly be a rational choice of frequency on the part of their designers. It was as if we were to leave our lighthouses flashing in broad daylight.

It was a plausible argument . . . but not, after all, entirely convincing. For some unknown reason the aliens may have found it useful to employ this frequency anyway. Maybe they liked things that way. We humans, after all, have done more foolish things than that. Hewish decided to perform an alternative test.

He decided to see whether the radio source was located on a planet. Any extraterrestrial civilization would be forced to exist on some planet orbiting about a star. It would be *in motion*—in orbital motion about its sun. And this motion was easy to detect. Hewish used the Doppler effect.

Although it may not be known by name, everyone is familiar with this effect. It operates every time a car drives by with the horn blaring. The abrupt decrease in tone as it passes comes about because the car, initially approaching, has suddenly begun to recede. It is this change in relative motion that is responsible for the change in frequency of the received sound waves.

What is true of sound waves is also true of radio signals from a pulsar. In this case, however, it is not their tone, but the *rate of pulsation* that would be altered. As the planet orbited about its sun it would alternately approach and recede from the Earth. Via the Doppler effect, this continually changing motion relative to us would induce a corresponding regular alteration in the detected rate of pulsation.

Hewish's test was negative. The most careful searches showed the alteration to be absent.

Bell: "Just before Christmas I went to see Tony Hewish about something and walked into a high-level conference about how to present these results. We did not really believe that we had picked up signals from another civilization, but obviously the idea had crossed our minds and we had no proof that it was an entirely natural radio emission. It is an interesting problem—if one thinks one may have detected life elsewhere in the universe, how does one announce the results responsibly? Who does one tell first?

"We did not solve the problem that afternoon, and I went home that evening very cross—here was I trying to get a Ph.D. out of a new technique, and some silly lot of little green men had to choose my aerial and my frequency to communicate with us. However, fortified by some supper, I returned to the lab that evening to do some more chart analysis. Shortly before the lab closed for the night, I was analyzing a recording of a completely different part of the sky, and in amongst a strong heavily modulated signal from [the radio source] Cassiopeia A . . . I thought I saw some scruff. I rapidly checked through previous recordings of that part of the sky, and on occasions there was scruff there. I had to get out of the lab before it locked for the night, knowing that the scruff would transmit in the early hours of the morning.

"So a few hours later I went out to the observatory. It was very cold, and something in our telescope-receiver system suffered drastic loss of gain in cold weather. Of course this was how it was! But by flicking switches, swearing at it, and breathing on it I got it to work properly for five minutes—the right five minutes on the right beam setting. This scruff too then showed itself to be a series of pulses, this time 1.2 seconds apart. I left the recording on Tony's desk and went off, much happier, for Christmas. It was very unlikely that two lots of little green men would both choose the same improbable frequency, and the same time, to try signaling to the same planet Earth."

Too bad.

"Over Christmas, Tony Hewish kindly kept the survey running for me, put fresh paper in the chart recorders, ink in the inkwells, and piled the charts, unanalyzed, on my desk. When I returned after the holiday I could not immediately find him, so I settled down to do some chart analysis. Soon, on the one piece of chart, an hour or so apart in right ascension, I saw *two* more lots of scruff [pulsars number three and four]. It was another fortnight or so before [the second] was confirmed, and soon after that the third and fourth were also. Meanwhile I had checked back through all my previous records (amounting to several miles) to see if there were any other bits of scruff that I had missed. This turned up a number of faintly possible candidates, but nothing as definite as the first four.

"At the end of January, the paper announcing the first pulsar was submitted to *Nature*. A few days before the paper was published, Tony Hewish gave a seminar in Cambridge to announce the results. Every astronomer in Cambridge, so it seemed, came to that seminar, and their interest and excitement gave me a first appreciation of the revolution we had started. In our paper we mentioned that at one stage we had thought the signals might be from another civilization. When the paper was published the press descended, and when they discovered a woman was involved they descended even faster. I had my photograph taken standing on a bank, sitting on a bank, standing on a bank examining bogus records, sitting on a bank examining bogus records; one of them even had me running down the bank waving my arms in the air—look happy, dear, you've just made a Discovery! (Archimedes doesn't know what he missed!) Meanwhile the journalists were asking relevant questions like was I taller than or not quite as tall as Princess Margaret (we have quaint units of measurement in Britain) and how many boyfriends did I have at a time?"

At about this time, Bell's part in the story came to an end. She stopped making observations and handed things over to the next generation of graduate students. She accepted a job in another part of the country in an entirely different field of research. She wrote up her thesis.

The pulsars went in an appendix.

On the ninth of February, 1968, the Cambridge group sent their paper announcing the discovery of pulsars to the British journal *Nature*. Its title was "Observation of a Rapidly Pulsating Radio Source," and it was published in two weeks—an unusually rapid decision on the part of the journal's editors, and one that testified to their sense of the importance of the discovery. That issue of *Nature* carried on its cover the words "possible neutron star."

Within a matter of weeks other research groups were getting into the act. There was a mad scramble to whip together the special equipment required to observe the pulsars. There was a rush to the telescopes. Astronomers who had been promised the use of a telescope for some entirely different project found themselves bombarded by telephone calls from colleagues anxious to use it for pulsar observations. Deals were made. A night on the telescope now was traded for a week six months later. And coincident with this effort, and parallel to it, was the continuing debate on the object responsible for the pulsating emissions.

Over and over again in this debate the same point kept recurring that had occupied Hewish and Bell. It was widely felt that the extreme regularity of pulsar emissions was something worth paying attention to. There was nothing unusual in the fact that they emitted radio waves. Many objects in the sky did this: our own Sun for example, albeit very weakly. Nor was it surprising that the signals fluctuated so rapidly. After all, quasars scintillated all the time—Hewish's telescope had been built to observe these fluctuations. But quasar scintillations were erratic. Before the discovery of pulsars no astronomical source had ever been found to pulse with anything approaching their regularity.

Clearly the pulsars contained within themselves a timekeeping mechanism, some sort of naturally occurring clock controlling their emission. It was on the structure of this clock that most attention focused. The debate on the ultimate nature of pulsars turned out in the end to hinge on just this question, and the way in which it was conducted is an example of the precise, abstract methods of science at its very best. It was a classic argument, beautiful in its sweep and generality, and its final resolution was one of the most telling illustrations I know of the power of abstract reasoning when combined with hard observations. Even now, more than a decade after the fact, it is an example that thrills.

For a variety of reasons it was found that no ordinary star could ever be a pulsar—not the Sun, not any star visible to the naked eye. Only small ultradense stars were capable of emitting such signals, and only two such stellar types existed. The first was the *white dwarf star*. White dwarfs are a fairly common stellar type, although they are so very small—about the size of a planet—that they are difficult to see. Not one is visible to the naked eye.

The second type of ultrasmall star was the neutron star. It had taken a regular train of radio bursts to bring them to the fore.

Three different clock mechanisms were proposed at one point or another during the debate on the nature of pulsars. The first we are all familiar with. It is the passage of the seasons. Winter is followed by spring which gives rise to summer and then fall, and the cycle continues—without end, and with great regularity. This ponderous cycle, this clock, is produced by *the Earth's orbital motion about the Sun.*

It is a far cry from the yearly orbit of the Earth to a series of bursts of radio emission a mere 1.3 seconds apart. How to bridge the gap? The proposal was that a pair of white dwarfs, or neutron stars, could be in orbit about each other. In fact such stellar pairs, or binary stars, are well known to astronomers. A binary system consists of two stars in close proximity to one another orbiting about their common center of mass. Some are visible to the naked eye. The second star in the handle of the Big Dipper is a binary, for example, and its two members, Mizar and Alcor, can be distinguished by those with good eyesight.

It is not so very difficult to imagine how such a pair could emit bursts of radio waves. Figure 1 shows one possibility. They could be generated at the point of contact of the two stars. Surely this is at least possible. After all, the stars are skimming by each other at enormous velocities. Furthermore, as seen from the Earth this emission would appear pulsed. The bulk of the two stars would hide the source from us during most of the orbital period.

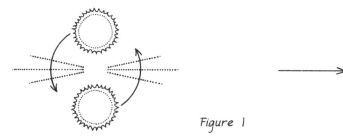

to the earth

Figure 1

There are only two configurations in which the point of contact would be visible to an observer stationed on the Earth: one is illustrated in Figure 1; the other occurs half a revolution later. Within such a model we would receive two brief bursts of radio emission for each orbital cycle of the pair.

A second possibility is illustrated in Figure 2. It relies on the prediction, discussed in Chapter 9, of Einstein's general theory of relativity that the gravitational fields of massive bodies bend the paths of light and radio waves. The bending is large if the bodies are very

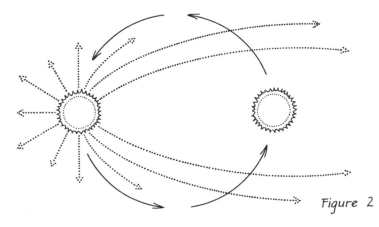

to the earth

Figure 2

dense: only white dwarfs and neutron stars are sufficiently compact to produce much of an effect. Each member of the binary pair would act as a lens, focusing the signals from the other. If one of the stars were to emit radio waves steadily and in all directions—not a very unusual state of affairs—the lens would bend their path and focus them into a beam. The beam would swing about as the pair revolved; as it swept by the Earth we would receive a pulse.

At least in principle, then, a binary star system is capable of emitting a regular train of radio bursts. Our own earthly astronomical clock ticks quite slowly. Can a much faster clock be imagined that ticks as rapidly as the pulsars?

It is a common feature of all binaries that the closer the two members lie to one another, the more rapidly do they swing about in their orbits. The very same principle is at work in the Solar System. The Earth orbits about the Sun once a year, while Mercury, closer in, completes its swing in 88 days. To construct a rapidly ticking clock we have to construct an exceedingly tight orbit. But there is a limit to the tightness that can be achieved. There could not possibly be a planet so close to the Sun that it orbited about it once a second. The closest an orbit can lie to the Sun is one that just barely skims along its surface, and in such a "minimum" orbit a planet would swing about the Sun once every three hours. Similarly, two ordinary stars can never orbit about each other any more rapidly than once an hour or so.

But white dwarf stars, being smaller than ordinary stars, can get closer to one another. A binary system consisting of two white dwarfs lying quite far apart would have an orbital period of one year—like the Earth. A somewhat tighter system can achieve orbital periods of an hour. But even here the members of the pair are still widely separated. In their closest configuration, in which the two white dwarfs swing about each other barely grazing their surfaces, so rapidly do they move that their "year" can be a mere one second long. And two neutron stars, smaller still, could whirl about each other 1,000 times per second.

So much for the first possible model of the pulsar clock. Orbits are not the only timekeeping mechanism that one finds in nature. In daily experience we have available to us an example of another: the twenty-four-hour cycle of day and night. It is produced *not by the orbit of the Earth but by its spin.* Can a model of pulsar emission be constructed based on rotation?

It is easy to see how such a model could produce pulses. We imagine that the radio emission is confined to a small region on the star— to a spot. Only when this spot is visible from the Earth would its emissions be visible; because the star is rotating it is regularly swinging into and out of view, as Figure 3 illustrates.

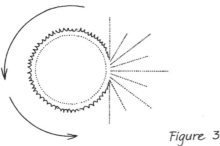

to the earth

Figure 3

Once again, the extreme rapidity of pulsar bursts puts severe restrictions on the model. It is one thing to spin some everyday object such as a basketball as rapidly as the pulsars tick; it is another matter altogether to spin something as huge as a planet or a star so fast. To illuminate the difficulty consider what would happen if one were to attempt to mimic the pulsars by accelerating the rotation of the *Earth* from its present rate of once every 24 hours to the required rate of once a second.

Do this by the imaginary experiment of mounting giant rocket engines sideways about the equator. On a signal they all commence firing. In response the length of the day grows shorter. The Sun rises and sets with increasing rapidity.

As the process continues things begin to grow lighter. Boulders that once were immovable can now be heaved about with ease. Overweight men and women gaze down upon the bathroom scales with satisfaction. The faster the Earth spins, the less things weigh.

Eventually what was once a matter of some amusement becomes more serious. People no longer can walk, but bound awkwardly down the sidewalks in enormous leaps. Their muscles are too strong for their weights—each person weighs a mere few ounces. A mild gust of wind is enough to send automobiles slithering sideways from their parking places. Stones jostle about alarmingly in the fields. As the rockets keep firing, a critical point is ultimately reached at which the day lasts just 1.4 hours. At this point every object along the equator weighs *nothing*. And as the rate of spin is increased still more, this area of weightlessness extends further north and south away from the equator.

Everything floats away into space. People drift helplessly upward. So do their automobiles. A vast jumble ambles off into the sky—animals, machines. Finally a positive force builds up, and it tugs upward upon everything remaining. Trees are uprooted and hurled into space. Vast chunks of ground tear loose and fly away. The very fabric of the planet is torn apart.

The Earth is ripped apart by rotation. This happens in exactly the same way and for the very same reason that a flywheel tears apart if set spinning too rapidly. Every object in the universe has a natural limit set upon its rate of spin, and if it rotates more rapidly than this limit it is destroyed. The Earth would be destroyed if it were to be spun more rapidly than once every 1.4 hours. The Sun flies apart more easily: it can only be spun once every 2.8 hours. If we are going to explain the pulsars on the basis of some rotating object, we had better be careful to search for one which at the very least is capable of spinning once a second.

As in the case of orbital motion, we are again thrown into the realm of ultrasmall stars. The more compact a star, the more rapidly it can spin. In fact it is gravitation that regulates this effect. The denser a star, the stronger is the force of gravity holding it together, and the more it can withstand the centrifugal effects of rotation. Only white dwarfs and neutron stars are sufficiently compressed to rotate at the required speeds. So strong is gravity on the surface of a white dwarf that a one-hundred-fifty-pound man there would weigh a full 25,000 tons, and on a neutron star gravity is stronger still. White dwarf stars, in consequence, can rotate several times a second; neutron stars, thousands of times a second.

The third and final model of the pulsar timekeeping mechanism is not based upon anything present in our common experience. But it is based on a phenomenon well known to astronomers—so very well known that it is probably fair to say that it was this model that first sprang to most astronomers' minds when the announcement of the discovery of pulsars came out. Certainly the discoverers themselves leaned toward this interpretation in their initial thinking.

It is *the vibration of a star*. Certain stars grow larger and then smaller in a regular progression. Figure 4 diagrams the motion: the star continually oscillates about, expanding and contracting without end, and as it does so it brightens and dims.

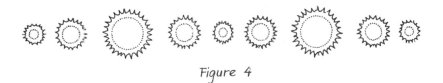

Figure 4

This regular variation in brightness can even be seen with the naked eye in a few cases. For example, the pole star vibrates, and while the vibrations themselves cannot be detected without the use of highly technical equipment, the associated brightening and dimming can. If one observes Polaris with the naked eye quite carefully, one will actually be able to see its changes in brightness.

An important feature of stellar vibrations is that they are quite without end. They do not die out. In this regard the vibrations of a star are very different from those of a bell, for instance, which rings only for a short while after having been struck by a hammer. Neither is the hammer blow required to set the star into vibration. It oscillates for reasons purely internal to itself.

Since vibrating stars vary in visible light, it would be no surprise to find them varying in radio emission as well. The signals might be emitted at some fixed point in the cycle of expansion and contraction of the star; for example, when it reached its minimum size and bounced outward. Alternatively they could be generated at an intermediate stage in the expansion, when the star was rushing outward most rapidly and ramming against its atmosphere, or corona.

Once again, the rapidity of the pulsar clock selects only certain possible candidate stars. It will come as no surprise to the reader to learn that only white dwarfs and neutron stars can vibrate as rapidly as the pulsars. The rate of vibration of a star is determined by the force of gravity upon its surface: the stronger this force, the faster it vibrates. In this regard, a star's rate of oscillation is similar to its maximum rate of rotation: both are fixed by the same physical quantity.

Ordinary stars like the Sun and Polaris have relatively weak gravity: Polaris oscillates once every four days, and the Sun, if for some reason it were to be set vibrating, would do so once every several hours. A white dwarf, on the other hand, could easily vibrate once a second, and in the case of a neutron star we have an embarrassment of riches: such stars, if they vibrate at all, do so thousands of times per second. So if the pulsar clock is provided by the vibration of a star, the star can only be a white dwarf.

The three models of the pulsar clock were proposed in the first few months after the announcement of their discovery, and they were debated fiercely. The pace of research was intense in those months. It was a heady time. A snowstorm, a positive blizzard of scientific papers whirled through the pages of the research journals. Papers were published announcing the discovery of new pulsars. Papers were published announcing the discovery of some new and previously unsuspected aspect of their emissions. Others described in glowing terms various theories of the mechanism responsible for these emissions— all mutually contradictory—and still others, equally contradictory, on the mechanism responsible for their regularity. A researcher would publish the results of his work. Almost before the ink was dry, he would encounter another paper containing some new point that he had neglected to consider. He would dash back to the drawing boards and in due course a new, or revised, paper would appear. It is no exaggeration to say that in the space of a mere few months, more research was done on neutron stars than in all the decades since they had originally been proposed in 1932.

The months rolled by. Confusion reigned, and as late as the fall of 1968, one year after the discovery of pulsars, the situation was still wide open. But then three things happened. Each was of decisive importance, and taken together they resolved the question completely. These three discoveries were made in as many consecutive months—October, November, and December—and almost before people realized what had happened, the problem was solved. By Christmas it was all over but the shouting.

The first was the discovery of the Vela Pulsar. The pulsar was located in the southern hemisphere constellation of Vela and it was very rapid—ten times as rapid as the others. It was almost *too* rapid. Its pulsation rate was nearly too great to be accounted for on the basis of any model involving white dwarf stars. Vela strained these theories to their limits. It forced them into unlikely extremes, into ad hoc assumptions, and it shifted the balance of probabilities toward the hypothesis that pulsars were neutron stars.

But far more important than this, *the pulsar lay within a supernova remnant.* Perhaps the reader will have anticipated this beforehand. A neutron star, born in the fires of a supernova explosion, invariably begins its existence embedded in the explosion's debris. If pulsars were neutron stars, they would have been accompanied by these remnants. But the first pulsars to be discovered were not. Hewish and his group had established conclusively that the original four were located nowhere near any known supernova remnants. The same was true of those discovered subsequently—until the Vela Pulsar came along. But the significance of this fact was unclear. At first glance it appeared to throw the weight of evidence in the direction of the white dwarf hypothesis, but upon more careful consideration it did no such thing—for supernova remnants did not last very long. Like the airborne debris of an ordinary explosion, they were evanescent. Perhaps the fact that the pulsars were not associated with any known supernova remnants merely meant that they were old.

The discovery of the Vela Pulsar established the first clear link between pulsars and neutron stars, and when taken together with the difficulty that white dwarf theories experienced in accounting for this object's rapidity, the evidence became still more compelling. It was beginning to appear that Jocelyn Bell had discovered the first neutron star.

Almost before the astronomical community had time to assimilate Vela's significance the second of the three seminal discoveries of 1968 was made—and if the first had been important, this was crucial. It broke the back of the problem. It was the discovery of the Crab Pulsar. This pulsar too was embedded in a supernova remnant, but its true significance lay elsewhere. It was in the pulsar's extreme rapidity. The Crab emitted at the unheard-of rate of 30 bursts per second, and if the white dwarf theories had been strained by Vela, they were annihilated by the Crab. This pulsar was too fast to allow the slightest ambiguity; its 30 bursts per second sounded the death knell for every idea involving white dwarf stars. With this one stroke, entire groups of theories tumbled into oblivion.

Only two possibilities then remained: rotation and orbital motion of neutron stars. How to decide between them?

The decision was made by the last of the three discoveries of 1968, but before recounting this discovery it will pay us to pause. There was an anomaly, a peculiarity in the facts of the matter as they stood in November of that year, and the anomaly was suggestive. Most of the pulsars in the sky bore no visible relation to supernova remnants. Only two, the Crab and Vela, were situated within such remnants. And these were the two fastest pulsars.

Were these facts related?

Why should it be that out of all the pulsars in the sky, only the two fastest occupied these remnants? By this stage it had become clear that pulsars were neutron stars, and that their births were marked by one of the most violent and cataclysmic events known to science: the supernova explosion of a star. The slow pulsars were no longer accompanied by the remnants of these explosions. They had outlived them. It was only the faster ones that were still surrounded by the debris of their birth.

It was almost as if the faster pulsars were the younger ones.

As indeed they were. A mere one month after the discovery of the Crab Pulsar, the group that found it observed it again—and observed that it was ticking more slowly. The Crab Pulsar was slowing down.

All at once the first inkling of an enormous vision rushed over the scientific community. In an instant the vast sweep of pulsar evolution was laid bare. A pulsar began its existence wrapped in the fires of the supernova explosion of a star. It began pulsing with great rapidity—far more rapidly than any pulsar known to us now. As the ages passed—thousands of years, millions of years—the pulsar steadily slowed, and as it slowed, its enveloping supernova remnant expanded. After one thousand years the pulsar had slowed to 30 pulsations per second—it had become the Crab Pulsar—and its surrounding remnant expanded ten light years—it had become the Crab Nebula. After 20,000 years it had evolved into the Velva Pulsar: pulsing more slowly, embedded in a nebula far more extended and tenuous. Ultimately, as the millennia rolled by and the remnant entirely dissipated into the vastness of interstellar space, the pulsar was stripped of the signs of its birth. Now it was pulsing a mere once a second.

The discovery of pulsar deceleration also nailed down their underlying nature, for of the two remaining models of the pulsar clock, only one was even capable of slowing down. This model was rotation. Spinning objects could easily decrease their rate of spin, and ultimately come to rest. Spinning tops did it all the time. But as for orbital motion just the opposite was true. These clocks could not slow down. Indeed, they sped up.

They sped up because of the emission of gravitational waves. Any orbiting object emits such waves. The Earth does so right now in its passage about the Sun, and so do the other planets of the Solar System and every binary star. The consequence of the emission of this gravitational radiation is that the orbiting object slowly spirals inward. Right now the Earth is microscopically drifting toward the Sun—and the closer it gets, the shorter is the year.

Within the Solar System this process is exceedingly weak. It is so minute that it has not the slightest practical significance. Over the entire history of the existence of our planet—more than 4 billion years—it has literally moved the Earth inward not even a hair's breadth and the year has not shortened by a second. But because they were orbiting so rapidly, this emission would be stronger in the case of the postulated binary system containing two neutron stars. Since they emitted more strongly, they would spiral in toward each other more rapidly—and the pulsar clock would speed up quite noticeably.

With this last small step the debate on the nature of the pulsars came to an end. The discovery of pulsar deceleration ruled out the binary star model, and by elimination only one model remained. Pulsars were *rotating neutron stars.*

So it was that neutron stars were discovered, and the ideas of Baade, Zwicky, and Landau borne out. They were borne out a full 33 years after these scientists had proposed them, and in ways they could not possibly have foreseen. It happened by accident, when a young woman discovered the first pulsar. Or perhaps it had happened earlier, when an eminent astronomer decided to build a new kind of radio telescope. Or perhaps it did not happen until a full year had passed and the Crab Pulsar and its deceleration was found, and their significance was forged in the hard fires of an international debate. It happened, at any rate.

In some ways it was an unsatisfactory discovery. For after all, what direct evidence have we found for the existence of neutron stars? The entire story is based upon a rapid train of radio pulsations whose rate is gradually diminishing. It is a long way from this to the actual observation of a tiny ball of neutronic matter lost in the vastness of space. It has been a purely negative argument that we have recounted. People drew up a list of possible candidates and eliminated all but one. At its heart the story of the discovery of neutron stars can be reduced to a single question: what else could the pulsars be?

There is only one reply—nothing. Nevertheless even now, more than a decade after the fact, we have *still* not succeeded in directly observing a neutron star. They are too small to be seen. If it were not for their strange ability to emit radio pulses, they never would have been found at all.

The discovery of neutron stars raised more questions than it answered, and showed them to be far stranger beasts than anyone had thought. Nothing in the original ideas of Baade, Zwicky, and Landau gave the slightest indication that they would behave in such a way. After all, the Earth rotates, and the Sun, but they do not emit pulses. Neutron stars, alone among every class of star in the sky, have this uncanny ability. This arose from their emission of radio signals from a small portion of their surfaces, but no one understood how they did this. No one knew how they acted for all the world like cosmic lighthouses. And no one knew how they slowed down.

3

Radio Telescope

The Quabbin Reservoir lies in the very heart of central Massachusetts, and it is surrounded by nearly 200 square miles of deserted watershed. This watershed is very nearly a wilderness. There is not a shopping plaza, not a town, not a single inhabited building to be found anywhere within it. It is closed to every form of commercial development and to recreational use such as hunting and picnicking as well, save for certain selected areas. Wild animals abound throughout the preserve. Pure streams babble amid deserted valleys. It is a lovely and unique area, and all the more so for being located within one of the most heavily populated regions of the entire United States.

The Quabbin Reservoir was flooded in the 1940's to provide water for Boston. Four towns were inundated in the process. Some 7,500 graves had to be dug up and reinterred out of reach of the rising waters. Homes were abandoned; some sold for their wood and others given away free to anyone who would guarantee their removal to safer places. Today the names of these vanished towns have a ghostly ring about them: Dana, Prescott, Enfield, Greenwich. None was ever more than a tiny hamlet—the largest, Greenwich, reached a population of some 1,500 souls in the year 1800, and it had been declining ever since. Old photographs of the region show quiet streets set about with wooden buildings of that austere white construction so characteristic of rural New England. Children played in broad fields; the stately homes were graced with wide porches; churches faced the town commons. Now all that is gone, and all that remains is an occasional gaping cellar hole to be found in the mud at low water, and the cracked ruin of an abandoned railroad.

Extending deep into the reservoir is a long finger of land, once a broad rise separating two valleys, and now, since the flooding of these valleys, the Prescott Peninsula. Halfway down the Prescott Peninsula, in a small clearing in the forest, stands a radio telescope.

This telescope was built under the direction of G. Richard Huguenin, an astronomer on the faculty of the University of Massachusetts, and I like to believe that the crusty and independent New Englanders displaced by the Quabbin Reservoir would have approved of the way in which he built it. Huguenin's great love is the building of scientific instruments, and when pulsars were discovered, he decided to build a telescope with which to study them. Had the climate of the times been different, he would have sought massive Federal funding for this purpose; and had he obtained these funds, his life thereafter would have been one long round of Big Science. Huguenin would have floundered in a maze of forms in triplicate, of contractors and subcontractors and labor disputes. But this was a time of cuts in the federal budget and the massive funding Huguenin needed was not available. He did not have enough money to build a radio telescope.

So he built one anyway.

Huguenin and his colleagues sat down at their drawing boards and produced a set of plans for a radio telescope specifically designed to be cheap. They designed it so that as many components as possible could be bought off the shelf in local stores. Visits to Sears took the place of visits to the National Science Foundation. Searches through lists of funding agencies gave way to searches through the classified ads. They got things on sale at hardware stores. They dug up an old used truck. They bought vast quantities of fencing wire from Sears and they used it, unorthodoxly, as the reflecting surface on their telescope. One of the engineers had a father-in-law who was a used car salesman, and he located for them an ancient telephone company vehicle equipped with an auger for drilling holes. They learned how to operate this auger, and with it they dug the holes in which the supports for the reflectors would be placed. As for these supports, they decided against a number of elegant possibilities and settled for the cheapest and most prosaic thing they could find—telephone poles. They erected those poles themselves. They cleared away the forest themselves. They laid telephone lines themselves. They scrounged and they hustled and they badgered: they got a local college to give them a computer, a corporation in Boston to give them a tiny control building, and everyone to give them their time.

The telescope that rose beside the ruins of Dana, Prescott, Enfield, and Greenwich was no grace to the landscape. It looked as out of place as a rusted jalopy in a leafy dell. It *was* rusted in parts, and it was set about with weeds. Four fencing-wire reflectors hung supported from telephone poles, and high over each was precariously dangled a radio feed. Coaxial cables running from the feed to the control building lay tangled in a rusty heap upon the ground.

Inside that building was rack upon rack of electronics, each festooned with knobs. They looked impressive, but not inordinately so —not so impressive as the cockpit of a commercial jetliner, for instance. Telescope operators lounged in chairs, sometimes studying the dials and sometimes not appearing to notice them. A stack of well-thumbed magazines lay upon a table: *Newsweek, Playboy*. Outside, visible through a picture window that looked out upon the clearing, deer grazed beneath the reflectors. And down upon this scene, down upon the nearby waters of the Quabbin Reservoir, down upon the bumpy dirt road leading up the Prescott Peninsula, down upon the deer and the very surfaces of the leaves that turned so glorious a color in the fall—continuously, steadily, and utterly invisibly there fell a rain of radio signals from the pulsars. It fell upon a rusty mass of metal in a clearing, and this mass alone among everything in the scene responded to it. Even as the deer grazed beneath its surface the telescope was in operation: automatically tracking the pulsars as they passed overhead, automatically gathering the data and storing it on computer tape. Once a week someone would toss the tape in the back seat of a car and bring it to the University of Massachusetts for analysis. And so Huguenin was listening to the pulsars.

Huguenin began work on this telescope when he moved to the University of Massachusetts in Amherst. Before that he had been at Harvard.

"At the time pulsars were discovered I was in the process of recovering from the loss of a satellite program," Huguenin told me. "I had been working for many years on a satellite designed to study the Sun at radio frequencies. We had designed it and flown parts of it on rockets and had the package essentially all ready to go when NASA's budget was cut. One of the easy places they found to save money was the university program because universities weren't very well organized down in Washington. We screamed the least. I was fairly young at the time and I had spent five years on the project altogether, so it was quite a blow. It had happened late in 1967 and it was just a couple of months after that that we got the airmail edition of *Nature*. I picked it up and saw the paper announcing the discovery of pulsars, and I immediately xeroxed it.

"It struck me as fascinating at the outset. Up to that point we had known about the slow variability of quasars—that quasars can vary their radio emissions over periods of years or even months, and even that was a real shocker. That something the size of a galaxy could vary its emission on the time scale of a month just didn't seem possible. And I guess we knew about various sorts of optical variability of stars, some of which vary their light on time scales such as days. But when something came along having *really* short time scales associated with it such as the pulsars—well, that seemed just incredible. That something at least the size of a star and that far away could possibly vary on such a time scale I found impossible.

"My colleague Joe Taylor was finishing up his Ph.D. thesis at Harvard then. We decided that this *Nature* article would be a fantastic thing to follow up on. Within a day or so we were deciding what we wanted to do. We talked about it and decided that the really crucial question would be polarization in terms of what these things actually were. So we started out to build an instrument to measure the polarization of pulsar radio emission that we would take down to NRAO—the National Radio Astronomy Observatory at Greenbank, West Virginia.

"I called up NRAO and said we wanted to observe these things. We had a lot of work to do because they didn't have anything down there in terms of the receiver equipment needed to do it. I had a group left over from the satellite project, and getting together we realized we already had a lot of equipment that we were able to use—magnetic tape recorders, fast chart recorders: that kind of stuff. So we put it together and in March, six weeks later, we were ready to go. We got on their 300-foot telescope and observed the pulsar and got some data on the polarization of individual pulses.

"We went on then to observe at NRAO quite regularly. Every few months Taylor and I would go down for a week or so. I remember particularly the first time we observed one pulsar. It was one of the closest pulsars and it was incredibly strong; so strong that we could see structure within its pulses, and this was something new at the time. I turned the speed on the chart recorder up to record it in as much detail as I could. I had it running fast and the pen on the chart started going faster and faster and started to glow cherry red. And the next thing I knew it burst into flames! The damn thing set all of our data on fire!

"The setup at this telescope was something to behold in those days. This was before they had their new control building and before the days of miniaturized electronics, and the upstairs was just jammed full of stuff. We brought with us a van with three or four six-foot racks of equipment when we went down there: it was a hell of a lot of equipment and they didn't have any place to put us but in the basement. But the telescope was controlled from upstairs." The 300-foot was a transit telescope—it lay on its back and pointed straight up, and was able to observe objects only as they passed by overhead.

"The people upstairs would tell us when a pulsar was about to pass over. We would get everything turned on and running at high speed and all set to record the data and then they would stamp on the floor as a signal at the right moment. Later on things got a little more sophisticated, but it was exciting back then because we were learning first things.

"At the same time we began a search for new pulsars and found the first one other than the original four that the group in England had discovered. I remember that coincided with the visit of the visiting committee. Tommy Gold was in from the Arecibo Observatory in Puerto Rico. They were very interested in getting into this pulsar bit down there and he had heard the rumor that was sort of floating around the cafeteria at NRAO that we had found a new pulsar. I said yes we had, and I explained how we had been searching. Gold was madder than hell that we would not be looking where his Arecibo telescope could see. My reasoning had been that we knew Arecibo was going to look and that it was a far bigger telescope than the one we were using at NRAO. On the other hand I knew that it was not capable of surveying more than a limited portion of the sky. So I thought we would start where they couldn't observe. Gold was mad because we had found this new pulsar and they couldn't study it.

"In September of '68 I moved from Harvard to the University of Massachusetts at Amherst. With me I brought my engineering team and all of the equipment. Joe Taylor was officially at Harvard but he was with us as much as he was there, and soon he moved out to join us permanently. It was clear that pulsars were fascinating objects and that they needed to be studied. So I decided that this was what I wanted to do. I decided to build a radio telescope at Amherst."

At the time of this writing, several hundred pulsars are known, only two of which emit visible light and can be studied with an ordinary telescope. The rest emit radio waves. The study of pulsars is therefore largely the business of radio astronomy. In fact the waves emitted by the pulsars are of precisely the same nature as those emitted by any radio station, and it takes the same sort of equipment to pick them up. Astronomers study the pulsars in the same way that they listen to the evening news while driving home from work: with a radio.

In principle any radio would do. The set in one's car would be perfectly capable of receiving pulsar signals if only they were stronger. But they are too far away. If a pulsar could be moved close to the Earth, every radio and television set in the world would receive its emissions. We would hear a sound very much like the ticking of a clock: a regular train of pulses, a uniform series of clicks.

Because pulsar signals are so faint we are forced to build more elegant radios with which to hear them—radio telescopes. The principles of construction of such telescopes are just the same as those employed in any other kind of radio. There are three essential components: the antenna, which detects the signal; the amplifier, which magnifies it into an electrical current of usable proportions; and the loudspeaker, which translates this electrical current into sound. Radio telescopes commonly employ only the first two of these stages and dispense with the third. The needs of science being what they are, it is most productive not to listen to the pulsars, but to record their signals in some more quantifiable form: on a chart recorder or an oscilloscope, or to feed them into a computer.

The first stage, the *antenna*, is a device for capturing the incoming signal. Often it is simply a wire. Radio signals are waves, and any wire bathed by these waves will develop an oscillating electrical current within it. It is this current with which the amplifier works. The bigger the antenna wire, the more electromagnetic waves it intercepts, and the larger the current it develops.

In an automobile this antenna is the aerial sticking out of the fender, or a fine wire embedded in the windshield. In household radios it escapes notice altogether: it is coiled somewhere inside. In both of these cases the antenna is small and unobtrusive. In a radio telescope, however, designed to capture such faint signals, it is huge, spectacular—by far the most impressive part. As the pupil of the eye dilates in a darkened room, as oversized antennas festoon the roofs in communities where the reception is bad, so the radio telescope sports a huge antenna. It can be 120 miles of wire strung up on posts, as was Hewish's. It can be dish-shaped, looking for all the world like a gigantic radar installation, such as the Arecibo radio telescope in Puerto Rico: 1,000 feet in diameter, a vast artificial crater carved out of the hills, floored with panels of perforated aluminum. Or it can be the largest radio observatory of them all: 27 dishes arranged in an enormous Y and spread over nearly 500 square miles of New Mexico desert, their signals delicately added together—the Very Large Array.

Properly speaking, these dishes are not the antennas at all. They are reflectors. Their function is to gather the incoming radio waves from a large area and to focus them back onto the actual antenna, which is referred to as the feed: a small and utterly unobtrusive device suspended above the reflecting dish. The reflector's function is analogous to that of the lens in a camera, which gathers the light incident upon it and focuses it onto the true detector, the film. But no matter how large the reflectors may be, the electrical current developed within the feed is minute. Hence the second stage in the operation of the radio: *the amplification of this current.*

In everyday radios the volume knob controls the degree of amplification, but in radio telescopes this nicety is dispensed with. Radio telescopes are either on or off, and when they are on they are on full blast. This component is the second major item of cost. Modern amplifiers as used for astronomical purposes are truly extraordinary things, and they are overwhelmingly superior to anything available on the commercial market. The task they must perform is so exacting that each amplifier is specifically designed and built to operate at a single frequency, rather than day-to-day amplifiers whose operating frequency is adjusted by tuning the set. In terms of the effort and ingenuity that goes into its construction, the amplifier of a radio telescope is every bit as impressive as the reflector. Of course, it does not look as impressive—a mere few racks of electronics.

Ordinary radio antennas are omnidirectional and respond to signals from all directions. Such a set should therefore receive every station within its range at once and would generate no more than a meaningless gabble. We avoid this problem by the prior agreement, enforced by the Federal government, that every station shall transmit waves of a different frequency than every other; and we select the station by adjusting the amplifier to operate at its frequency—by tuning it. But celestial radio sources such as the pulsars transmit at *every* frequency of the radio spectrum. With them tuning the receiver would do no good. So the radio astronomer designs his antenna to receive signals only from a specified direction in the sky: to be not omnidirectional but unidirectional. You do not tune a radio telescope. You point it.

So we build a radio—any radio, from the smallest transistorized model that we find upon the beach to the largest and most sensitive of them all. Each in its way is a marvel of technology, and each exists for the purpose of putting us in contact with signals from a distant source.

The contact is established through the medium of the electromagnetic field. This field pervades all of space. It surrounds each and every one of us. It penetrates into the interiors of our bodies and down to the very core of the Earth. And it extends far out into the icy cold of interstellar space. So far as we are able to determine, there is not a single region of the universe free of this field.

It is never wholly still. The field quivers and ripples incessantly. Waves flow along it. Here is a faint ripple. It registers upon the retinas of our eyes as the light of a distant star. Here now is another ripple: light from a second star. Uncounted thousands of such waves flow into the Earth, and through them we see the nighttime sky: planets, stars, constellations. Here again is another wave, enormously stronger than the rest: it is sunlight. Then myriads more, but enormously weaker. Amplify them with a telescope and they reveal still more stars but also something new: galaxies, nebulae, and quasars.

Here now is a wave not weaker or stronger, but *longer* than the rest. Visible light is a wave in the electromagnetic field of a very short length—about a hundred-thousandth of an inch. Waves of either greater or shorter length we do not perceive as light. We do not perceive them at all; rather, we build instruments to detect them. Shorter waves are X rays, with a wavelength of a hundred-millionth of an inch, or gamma rays, with a wavelength of a hundred-billionth of an inch. Longer waves are infrared light or radio waves. Radio signals can have a wavelength ranging from fractions of an inch to miles. Radio stations emit such waves. So do television stations.

And so do the pulsars. They travel through "empty" space and impinge upon the Earth. Scattered over the Earth are radio telescopes. They listen.

Huguenin came out to Amherst with a plan in mind and a promise of $25,000 in seed money from the University of Massachusetts. The first job was to find more money. "That period around '68–'69 there were budget cuts in the Federal government," he told me, "and it was hard to get funds. In fact, it was just those years that the National Science Foundation closed down big operations at three major universities. For that reason we didn't look too favorably on the NSF. Similarly, NASA was out too. So we set out to raise money from small, nongovernmental foundations. We submitted a proposal to the Research Corporation, and also split it up and submitted it to the Sloan Foundation and the Fleischman Foundation. Much to our amazement all three responded, and we ended up with twice the money we thought we needed to get started."

I asked Huguenin what the criteria were in selecting the site for his observatory.

"First, it had to be flat," he replied. "Reasonably flat, not absolutely so. It made the job of building somewhat easier. But much more important than this was that it had to be as far away as possible from outside interference."

It is all too easy for a radio telescope to be swamped by man-made radio interference. After all, the signals it is attempting to detect are far weaker than the ever-present jamming from local sources such as radio stations and the like. There is a similar difficulty with optical telescopes. The great 100-inch optical telescope on Mount Wilson, in California, for years the biggest telescope in the world, is now entirely unusable, for it directly overlooks the city of Los Angeles. The neon signs and street lights down below make a pretty sight from the observatory grounds, but they ruin the nighttime sky. Similarly, the 200-inch telescope on Mount Palomar is too close to San Diego for comfort, and in recent years its effectiveness has been seriously degraded by the growth of that city.

"Highways and power lines are the main source of interference for radio astronomers," Huguenin told me. "In automobiles it is the sparking mechanism: the spark plugs, the wires to the plugs, the distributor and the coil and all that. They produce a tremendous signal and they can cause radio astronomers a tremendous amount of grief.

"Also, power lines are noisy. It is the insulators, primarily. What happens is that they get dirty or they crack and there is an arc discharge which radiates radio signals. You can also *see* it: it is most visible if the air is salty, and if you walk along the seashore at night you can see the discharge on almost every pole. I had had a lot of experience with this on a previous project. We had a truck back then with a little antenna and we could find out which power pole the interference was coming from and we would drive up to it in the truck. We would very gently drive the truck right into the pole and bang it. The shock knocked the crust of dirt off the insulator and then we were OK. We used to go around doing this at the National Radio Astronomy Observatory at Greenbank and people thought we were plain crazy.

"Radio stations can also be a problem. In Massachusetts you just can't get very far away from them because they are all over the place. But you can get away from them *in frequency*. In Amherst there is a station that puts out a strong signal at a frequency of 89.5

megahertz. So you don't observe at 89.5. You build your receiver to operate at 87 megahertz and then you are all right. At one point a new radio station came on the air at our operating frequency and we had to redo our whole setup. You move around."

Just as it requires a continual effort to prevent the few remaining wilderness areas of America from being overrun by shopping plazas, housing developments and the like, so the observing frequencies of radio astronomy are in need of protection. There is a very real concern that astronomy within the continental United States will be wiped out in the foreseeable future unless more stringent conservation measures are imposed. "There are laws and so on," said Huguenin, "but they don't do much good. In fact, there is a whole person at the National Science Foundation in the Astronomy section, called the spectrum manager, whose job is to watch over these things. There is a complicated set of regulations that are designed to protect radio astronomy but they don't work very well. There is only one absolute protection and that is a small range of frequencies around the 21-centimeter wavelength at which interstellar hydrogen emits. Nobody anywhere in the world can transmit in a band around that. But that was around the last chance we had to rope off a chunk of the radio spectrum for astronomy. Since then we have only been able to preserve other frequencies with partial protections but they are almost worthless. Surprisingly, the places that work best are the places that are exclusively assigned to somebody else. One of the frequencies at which we observe is the military aircraft frequency range. That frequency is extremely quiet except for the occasional time when the military does use it, in which case it wipes us out totally. Of course the reason this frequency is safe is no one wants to get on and run the risk of interfering with the Strategic Air Command: they know they are going to be come down on pretty hard.

"Anyway, we set about choosing a site. I bought a set of all of the topographic maps of the Commonwealth of Massachusetts. They make a large stack, and we went about searching for what looked like suitable land. We found two possibilities. The first was on the Prescott Peninsula, the second down on the Swift River. Both met all our criteria.

"That was in '69, in March. We put together a bunch of equipment which we carried around in a jeep to survey the two sites. I remember getting stuck in the mud and there was still snow on the road. It was actually easier to get access to the Swift River site, so that was our first choice but the tests there showed that it was not good—there was too much radio interference. So then we went out on the Prescott Peninsula. We drove around and around the Peninsula with a wildlife guy from the University of Massachusetts who knew that place like the back of his hand. That was a real experience, exploring the area with him. There is a whole network of dirt roads out there. We finally came to the place where we are now and I knew that was it. It was a good location: fairly open, not a lot of trees to cut down, and level. We were monitoring the radio interference and it looked very good.

"The Quabbin site is an outstanding location. It is unique. There is hardly another place like it anywhere in the East—including most of Maine, although I think you might be able to go to northern Maine and accomplish the same thing there. But we were especially lucky in that this place was only 20 miles away from our home base in the University of Massachusetts at Amherst.

"By the fall of '69 we had obtained permission from the State to erect a telescope on public land and had amassed enough money to begin. So we started construction."

"We began by buying a chain saw on sale and started felling trees. We were out there cutting them down and clearing off the fields and getting ready to lay things out. Most of the wood was ash and it was carted off as firewood—the University got it, and it was used around the lodges and dormitories. We had borrowed a University pickup truck and everyday when we went back we took with us a truckload of wood."

The Quabbin observatory consists of four separate antennas whose outputs are precisely added together before being fed into the amplifier. "The reflecting surface of our antennas is a wire mesh," Huguenin told me. "Sort of a glorified chicken wire. It is used in lots of different ways—by farmers for fences and as windows on chicken coops, and in construction for reinforcement in the building of concrete embankments. 'Hardware cloth' is another name for it. Most of it we got from Sears Roebuck out of the catalog. That turned out to be the best. As time went on I got concerned that it might not last so long, and we found a company that made the stuff with twice as much rust-proofing. It cost 20% more but I thought it would be worth it. That turned out to be a mistake. If you go out to Quabbin today you will see that that is the stuff that is rusted."

Huguenin wanted his reflectors to be immovable. He wanted them to lie flat on their backs and point up to the sky, the reason being that such fixed structures would be relatively cheap. Each would therefore correctly focus pulsar signals if the pulsar lay directly overhead. The feed would be located at the focus point, as in Figure 5.

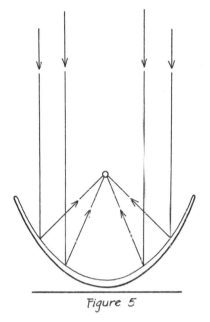

Figure 5

The problem was that such a reflector would not focus incoming signals from anywhere else. If the pulsar were off to one side of the vertical, each ray would be reflected differently, and they would never converge as in Figure 6. It was no good. Huguenin wanted a telescope with more flexibility.

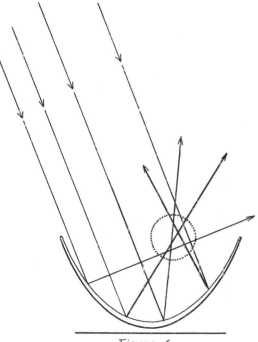

Figure 6

The straightforward way to deal with this problem was to make the reflector steerable (Figure 7). This configuration, like the first and unlike the second, did possess a focus point and would be capable of observing pulsars no matter where they lay in the sky. But it cost too much.

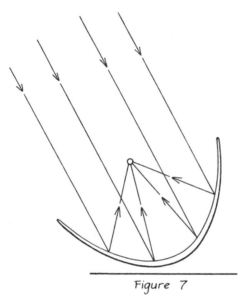

Figure 7

Huguenin adopted a compromise. Only if a reflector were shaped like a *parabola* would it focus incoming waves perfectly. Steerable radio telescopes therefore were parabolic in shape, as were the mirrors of optical telescopes. Reflectors of other shapes, on the other hand, focused imperfectly and as a consequence were not used very much. But a reflector whose outline was *a segment of a sphere* had a very interesting property, and one that made it useful to Huguenin. If the radio source lay directly overhead, a spherical reflector focused the incoming signals not as well as a parabolic dish, but not too badly for all that (Figure 8). But this partial drawback was to be set against the focusing properties of the reflector for a source lying off to one side of the vertical, illustrated in Figure 9.

The convergence was still fairly good. If the feed were to be moved off to one side into the region of partial focusing, it would still detect the incoming signals. The design would do.

Paradoxically, such a reflector would focus the signals from pulsars lying at many different points of the sky simultaneously. If there

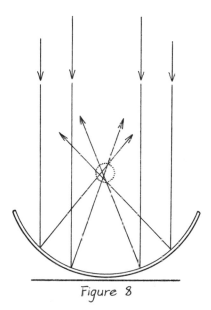

Figure 8

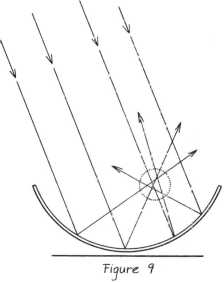

Figure 9

happened to be, say, four lying within its general field of view, there would be four different regions above the reflector in which pulsar signals were concentrated. Within each focus region the emissions from only one of these four sources would be compressed, and if the feed were placed there, it would detect that and only that pulsar. So Huguenin would point his telescope: not by tilting the reflector, but by moving the feed.

"Each of our reflectors is held up by twenty-two telephone poles around the outside edge," Huguenin told me, "the feeds by three bigger ones in the middle. New telephone poles. I knew we wanted to build the dishes as big as we reasonably could because of the cost of the amplification. The three tall poles determined the size of the entire dish and set their cost. They came from the Pacific Northwest —and *their* size was determined by the size of the largest thing you could get on a railroad car. They were 70 feet long. It turned out that if we used them we could have a dish 120 feet in diameter. So that is what they are.

"The poles were cheap because the power company uses millions of them. Likewise, the cable that holds up the reflecting surface is the same cable they use for guying. The mesh is fastened onto the cable with a hog ringer, which is a device used in the automobile upholstery industry. Everything was done by me and Joe Taylor and graduate students and some undergraduates: English majors and history majors as well as astronomy majors. We ended up building each dish for about $10,000 in equipment and $20,000 in labor.

"As for the phone poles, we put them up ourselves. It all went fine with the small ones. But our truck was not big enough to lift the big ones. You have to grab a pole above its center of gravity and these things were 70 feet long. It was clear when we first tried to lift one of the big poles that we were going to need some weight on the end as ballast in order to swing that end down into its hole.

"The first time we tried it we used *people* as ballast. That turned out to be a mistake. There is actually a movie of what happened. One of our engineers thought he would get a home movie of the operation, so he stood alongside filming it as the boom lifted the first pole. His movie shows the boom beginning to lift the pole and hanging onto one end are Taylor and about four other people. They are sitting on the pole like cowboys—riding it with their legs wrapped around the end. The movie shows the thing coming up with all these people on it, and the pole comes up some more . . . and then all of a sudden you see the pole start to rotate! It slowly begins spinning upwards, and all these guys going up into the air . . . and that's the end of the movie. The engineer taking the movie was the only one around who could keep the whole thing and all those people from going right up. So he dropped the camera, rushed off, and grabbed hold.

"After that we used a big hunk of steel for ballast.

"That winter we started the heavy fabrication. We got a surplus mess tent and pitched it and all winter long we used it for building the steels. We bought us a saw at Sears. All we had was just raw steel and we had to cut it to length and weld it together. There was the worst damn weather. We had an oil-fired torpedo heater: it looks like an oil burner but there is no burner; just a flame and it blasts out heat and smoke and stink and everything else—a big blowtorch. All winter long: just cutting steel and getting it stacked up.

"In the meantime we needed to have a place to put the electronics. So we bought a house trailer. I remember vividly the night they delivered that trailer—our long-awaited lab building. It was the middle of winter and we had gotten an ice storm. It was terribly slippery, and a guy showed up with this big 60-foot-long trailer and just a tiny little truck pulling it. He didn't have any chains. I said, 'You had better get some chains because the road out there is sheer ice.' 'No, I never get stuck with this truck,' he told me. At that time I had a Bronco—a jeep vehicle—and I led him down the dirt road out to the site.

"It all went OK till we got to a hill. The first part of that hill he handled OK, but then about halfway up the truck stopped: stalled, and with that big trailer behind it. He decided he was going to back down the hill to get another start. But as soon as he started the truck started to slip—and the next thing I knew the whole thing jackknifed in the road. That great big trailer right across the road.

"I was ahead of him and there was no way to maneuver around him. I couldn't get out. But I vaguely remembered from the time we had been out there the year or two before that somewhere on the Peninsula there was another road. Of course it hadn't been plowed. But I got my four-wheel drive engaged and he got in with me and we found that road. I got around the trailer and took him back to the University. We went over to a local garage to get a wrecker. I said, 'It is really slippery out there, so for heaven's sake bring chains.' 'Don't tell me how to do my business,' was the reply. We went back out with the wrecker and managed to get around the jackknifed trailer. It was so slippery up there that the wrecker, without pulling anything, just barely got up that hill. But he had a device that could dig in and anchor the wrecker. And then he ran out the cable.

"By now it was about 7 o'clock in the evening. The wrecker would tow the trailer and truck up 100 feet. Then with my Bronco I would pull the wrecker up another 100 feet. Then he would anchor himself again and he would pull the trailer up again and then I would pull the wrecker up again. We got to the steepest part of the hill. It was a combination, a god-damned railroad train with the Bronco and the wrecker and the tractor and the trailer going up the road—just barely making it with all of this ice.

"Finally we got up to the top of the hill. I said, 'I am not going down this hill in front of all that.' So we left it there and went back home. Next week we had the University come in with a bulldozer and bring it down that icy hill. A long story . . . but that was a very long night. It was midnight by the time we got to the top of the hill."

They implanted the phone poles, and from them they suspended cables. Over the cables they fastened the reflecting wire mesh and adjusted its shape through the simple expedient of dangling bricks beneath it at strategic locations. They suspended the feed from the extralong poles in the center, and strung coaxial cable from it down to the trailer now safely parked alongside. Within this trailer went the electronics for amplification, and a computer and various other pieces of equipment. By December of 1970 the telescope was ready to go and the first observations were performed.

They had spent $146,000. For such a sum one might buy a large sailing yacht, or an apartment in New York City. The United States Department of Defense runs through hundreds of billions of dollars every year: in about the length of time one can comfortably hold one's breath it spends enough money to build another Quabbin observatory.

"There are a number of reasons why Quabbin was so cheap," Huguenin told me. "Partially it is because the telescope was specifically designed with pulsars in mind, and this enormously simplified the task. But I would say the real reason it was so cheap was that it had to be that cheap. We didn't have any more money and we wanted to do it. I am sure that if we had twice as much money we would have spent it all: things would have gone a little faster and the telescope would have been a little fancier.

"That is the money we had to do it with," he said. "So that is what we did."

The dirt road leading to the observatory at Quabbin branches off from a small state highway skirting the watershed, and it is entirely unmarked. Nothing distinguishes it from any of the other access roads leading into the region. Turning onto it one day I was soon halted by a locked gate, and I opened it with a key I had picked up from the Astronomy Department offices at the University of Massachusetts. Passing through, I carefully locked the gate behind me.

It was late in the afternoon. The season was early winter but the first snows had not yet fallen. All the leaves were off the trees. The Quabbin wilderness held that grim and austere beauty so characteristic of New England in that season. The forest stood silent and still. Overhead, pale sunlight filtered down from a gray sky. Sounds and colors were muted. The trees stood gray and bare above a forest floor with a thick blanket of brown leaves. Crunching some of them in my hand, I inhaled their wonderful odor.

Driving down the dirt road, I passed innumerable stone fences in the forest. Long ago they had been piled up as settlers cleared the land, but now they were overrun with bushes and weeds and wandered aimlessly among the woods. The road skirted a muddy pond. Briefly through the trees I caught a glimpse of the great waters of the reservoir itself. Mile after mile rolled by. A startled deer leaped in front of the car, bounded across the road, and vanished among the trees. Halting, I caught sight of several others silently gazing at me, utterly still.

Arriving at last at the observatory, I found it a scene of bustling activity, in sharp contrast to its surroundings. It had expanded a good deal since it first went into operation. The old trailer stood abandoned now, replaced by a pleasant control building. There was running water and a phone. Standing beside the four pulsar dishes was a newer and sleeker telescope housed in a glistening white geodesic dome and dedicated to the study of molecules in interstellar clouds.

The pulsar dishes, in contrast to it, looked shabby. They were rusted in spots, and there were occasional holes in the mesh. Weeds grew underneath them. To the casual observer they looked abandoned. But they were not abandoned. Even as I wandered beneath them they were in operation. The system was totally automated now. It was a computer that gathered the data, and it did so entirely without human intervention: daytime, nighttime, holidays, cloudy weather, rainstorms—it ran continuously. The coaxial cables from the four antennas leading into the control building lay all tangled together in a heap. One might come across such a pile at an abandoned farm. Here too appearances were deceptive, however, for even as I watched, pulsar signals were flowing through them. Their lengths had been so carefully adjusted that the signals from each of the four antennas arrived at the building precisely in phase with those from all the others.

Inside, nothing was unkempt. There was a small and pleasant library containing reference books, copies of the latest journals, and a photographic atlas of the sky. There was a Xerox machine. There was a kitchen, and two tiny windowless bedrooms where observers who had been up all night could sleep during the day—or vice versa. There was a machine shop and an electronics lab. Walls were festooned with innumerable photographs of the telescope.

In the control room racks of electronics lined the walls; some fairly bristling with buttons, knobs, and lights, others entirely unadorned. (Over one was set a plaque on which was formally engraved RUSTIC YET POTENT.) There was an amplifier of extraordinary capabilities; an atomic clock to determine the precise instant at which each pulse arrived; an analog-to-digital converter, which took the output of the amplifier and translated it into a number with which the computer would subsequently deal. There was a typewriter used for communication with this computer.

The computer itself occupied a small rack all its own. It carried out four quite separate functions, and it did these simultaneously. It guided the four feeds, continually tracking the drifting focus points of the pulsar as it passed overhead in the daily east-to-west rotation of the sky. It monitored the status of the telescope for calibration purposes. It stored the results of the observations as they emerged from the electronics. Finally, after a pulsar had been studied enough it moved the feeds to observe the next one that had swung into view. At the time of my visit the observing program included 20 pulsars and called for roughly two-hour observations of each: every two days the computer cycled once through its list.

The net result of the entire operation was a reel of magnetic tape. It was mounted on a tape drive controlled by the computer and, even as I watched, data were being written onto it. Inch by inch it slowly advanced across the recordings heads. After it had wound completely through, an operator would rewind it, replace it with a blank, and bring it down to the University.

Standing before all this equipment, I was suddenly struck that nowhere in the room did I see any evidence of a pulsar. I did not hear a steady series of clicks from some loudspeaker connected to the output of the amplifier. I did not see a series of pulses drawn on an unwinding sheet of graph paper. The entire installation was dedicated to the study of pulsars but nothing in the room revealed one to me.

In this regard radio telescopes are far different from their more familiar optical counterparts. After all, one can *look* through an optical telescope. But more and more, nowadays, complicated and sophisticated electronic equipment is being mounted on the eyepieces of these telescopes too, and automated analysis is rapidly replacing the human act of looking. As its methods change, astronomy is losing direct contact with its subject matter.

As I stood there in the control room of the observatory at Quabbin, the complexity and the fundamental abstractness of the act of observing pulsars revealed itself to me. No one had ever seen a pulsar. No one ever would. I had come to the radio observatory looking for one—but all I found was the observatory itself. I was reminded of a definition of astronomy I had once heard: *astronomy is the study of telescopes.*

One cold winter day in December of 1980, Joe Taylor rewound the tape containing the results of that week's observations. As he had done so many times before, he removed it from the tape drive. But this time he did not replace it with a fresh one. Instead, he reached down and turned off the computer. He turned off the amplifier, and the analog-to-digital converter, and the atomic clock, and all the other equipment in the room. Lastly, he shut down the power for the entire operation.

After ten years of operation, the pulsar observatory at Quabbin had outlived its usefulness.

The group that had constructed it was breaking up. Taylor had accepted a job offer from another university and was preparing to leave. Huguenin was devoting all his energies now to the newer radio telescope standing alongside the old dishes. Most of the engineers and technicians who had built the pulsar operation had long since transferred their attention to this new device.

More and more in the last few years Huguenin, Taylor, and their students had been finding themselves journeying off to the larger observatories such as NRAO, Arecibo, or the Very Large Array to conduct their observations. More and more they had been realizing that the telescope at Quabbin had done everything that it was capable of doing. After all, it was relatively small. Rather than expand it by adding still more dishes, they decided to shut it down.

The computer went to the University of Massachusetts. Some of the electronics were transferred to the new molecular astronomy telescope, others to Taylor's ongoing program of observing sessions at the larger outside observatories. Technicians began removing the feeds and disassembling the reflecting dishes beneath. They resurrected the ancient phone company truck, and removed the telephone poles.

True to the spirit that had guided them from the beginning, Huguenin and Taylor looked about for a buyer for those used telephone poles. They found one. "We sold those poles," Taylor told me with a gleam in his eyes, "for more than we had paid for them in the first place."

4

The Electromagnetic Storm

By the end of 1968 it was generally understood that pulsars were rotating neutron stars. But this was all that was understood, short of the vague impression that the pulsations had something to do with the emission of radio waves from a specified location on the star. Neither was it known how they went about slowing down. The task of this chapter is to explain these things. But we will not be able to get very far. No one can explain them.

Before asking why neutron stars do what they do, it would be well to find out *what* they do. That is what radio telescopes are for. They are telling, in the richest possible detail, just what pulsar emission is really like. There is no way to understand the pulsars without assimilating this detail, for their emission is exceedingly complex. To say that it consists of a series of pulses totally fails to do justice to the rich and baffling structure exhibited by these pulses. Every facet of this structure is a potential clue.

The crudest instrument with which pulsar signals are studied is the chart recorder, which simply and graphically displays the output of a radio telescope. Figure 10 is a brief section of such a record for a typical pulsar. The first thing to be noticed in this figure is the base line—the interval between pulses. It shows some structure. But this weak emission does not come from the pulsar itself. It is noise: an amalgam of human interference, signals from cosmic sources other

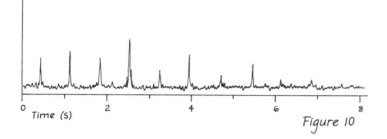

Time (s)

Figure 10

than the pulsar, signals from our own atmosphere, and even spurious signals generated by the radio telescope itself. Much of the effort of building finer telescopes is directed at reducing just this noise; and as this effort proceeds it becomes ever more clear that pulsars are totally quiescent between their bursts of emission. They are either on or off, and when they are off they are completely off.

As for the pulses themselves, they are very brief, far briefer than the interval of time separating them—and even this is a problem, for it shows that any model in which pulsar emission is simply thought to arise from a spot must be wrong. Such a picture is not capable of accounting for brief pulses at all. It yields broad pulses, in which the emission occupies just half the pulsar cycle.

In Figure 11 is sketched the rotating neutron star as it would be seen from the Earth, and upon it is indicated the localized spot from

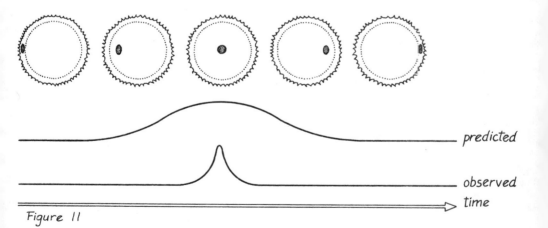

predicted

observed

time

Figure 11

which the radio emission is supposed to arise. As the star spins, this spot rotates about, alternately swinging into and out of our view. Figure 11 shows a series of "snapshots" of the configuration, arranged in chronological order with the time advancing from left to right, and below them the received radiation intensity—first, that predicted by the model; below, that observed.

In the first snapshot the radio-emitting spot has just appeared over the left-hand horizon of the star, and at this moment we begin to receive its emissions—but faintly, for the spot is pointing off to the left. As time passes, however, it swings more and more toward the Earth. In the second snapshot it has advanced partially across the visible disc of the star, and points more nearly in our direction. The intensity of the emission that we receive has increased. By the third snapshot the spot has reached the center of the star as seen from the Earth and the received intensity is at a maximum. Then, as still more time passes and the star continues to spin, the spot is carried still further and, in the fifth snapshot, the intensity drops to zero.

By this point the star has completed just half a rotation, and throughout all this time radio signals have been received from it. But this is not what is observed. The observations reveal a far sharper pulse, a brief burst of radio emission confined to a small fraction of the pulsar cycle. The model predicts too broad a pulse.

Where has the model gone wrong? It went wrong because of an assumption that we have been making—an assumption so natural and so obvious we never even bothered to state it. *We have been assuming the spot emits in all directions at once.* Figure 12 shows the radiation pattern from the spot we have been unconsciously assuming. The emission fills just half the diagram, and as the star

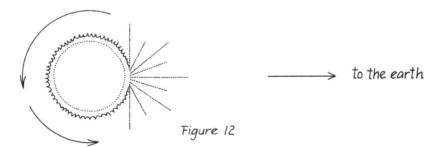

to the earth

Figure 12

spins emission will be observed for just half the time. It is an entirely reasonable pattern to assume. But it cannot be correct.

The observed sharpness of the pulse implies that *the radio emission from a neutron star is collimated into a beacon,* as shown in Figure 13.

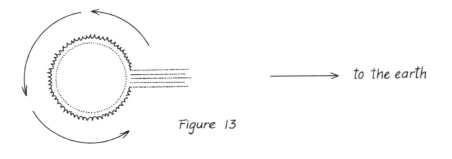

Figure 13

Only such a tight emission pattern is capable of yielding the observed brief pulses. With the discovery of this narrow beam, astrophysics enters an entirely unfamiliar realm. Nothing else in the astronomical universe is known to do such a thing. In some way, by purely natural processes, a neutron star acts as a kind of cosmic lighthouse.

Return now to Figure 10. In this figure successive bursts are of differing intensities. The pulsar is erratic in its strength: some pulses are quite bright, others so dim as to be nearly indistinguishable from the noise. Each succeeding pulse differs in its intensity from both the one that preceded it and the one to follow. The beacon fluctuates as it spins.

On other occasions a pulsar will exhibit a more extreme version of this fluctuation. Figure 14 shows a chart record in which the radio emissions have abruptly and completely ceased. This pulsar has quite

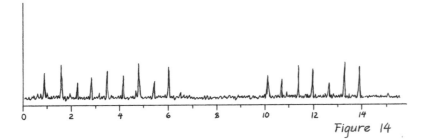

Figure 14

literally turned itself off for a time. The phenomenon is known as *nulling*, and perhaps the most remarkable thing about it is that it comes completely without warning. Nothing in the emission prior to a null gives any warning that it is about to occur. The beacon does not slowly fade to invisibility—it suddenly and completely vanishes. Such abrupt fluctuations are relatively common in the pulsars, and they last from a few seconds to a matter of minutes.

Figure 15 passes on to a higher level of complexity and studies the characteristics of the individual pulses in more detail. In this figure successive pulses have been stacked one on top of the other, and the "dead time"—the interval between pulses—removed for clarity. One reads the figure with time advancing from left to right across each

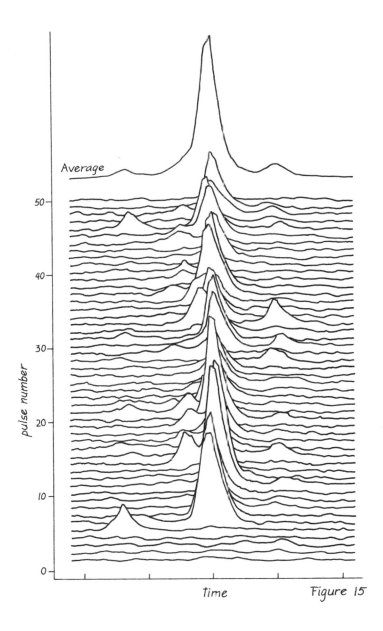

Figure 15

individual pulse, and from bottom to top from one pulse to the next. The first pulse in the data is represented by the bottom line, the second by the second line, and so on up the page.

In this stretch of data the first four pulses are simply missing: at the beginning of the observation the pulsar had been in a null state. The first pulse actually to appear in the data is the fifth, and it is somewhat weaker than the average. Not only is it weaker, but it arrived *earlier* than it ought to have—by some 0.03 of a second— and it is not until the sixth pulse that one comes in at the "correct" time and intensity. But everywhere throughout this diagram there is enormous variation from pulse to pulse. Not one of them is precisely like the others. Many, like the anomalous fifth pulse, arrive at the "wrong" time, and some others actually split into several components, the one too early, the other too late.

At the top of Figure 15 is represented, not the last pulse in the span of data, but something entirely different: the *average* of all the pulses contained in the data. And the striking thing about this average is that it is stable. Although individual pulses vary radically from one to the other, the average pulse shape never changes. Perhaps ten minutes worth of data are required to obtain such an average, but once found it is identical to that from any other ten-minute stretch of data—even one obtained several years before. The pulsar searchlight beacon grows stronger and weaker, it puts out fingers on either side and changes its detailed pattern—but never at random. It always remembers its proper shape.

This average pulse shape does not refer to the *most common* among the individual pulse shapes revealed in the data. As a matter of fact it does not refer to any of them. One can search among all the pulses shown in Figure 15 for a single one resembling the average in every detail, but the search will be in vain. The average pulse shape bears the same relation to individual pulses that the mythical average American does to real men and women. It is a statistical property of an assemblage.

Perhaps the most remarkable feature of these average shapes is that they are *different* from pulsar to pulsar. No two pulsars have exactly the same average shape; each shape, in turn, is unique to its own source. It is so unique that it can be used to identify the source—a fingerprint. Figure 16 shows a set of representative average pulse shapes for six pulsars chosen at random. Each is different from the others; some in quite obvious ways, others only on close examination. The first, for example, consists of a main pulse followed by a weaker one. The next two have relatively simple profiles, although the third is briefer than the second. But the fourth average pulse shape is quite complex: at first glance it appears to consist of

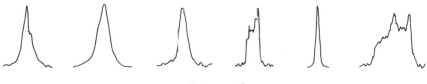

Figure 16

a subsidiary and then a main component, but upon closer examination each of these two components itself is seen to be composed of still finer structure. The fifth pulsar illustrated is again quite simple, but then, in the last, is found an average pulsar beacon consisting of fully five separate components.

We can also speak of *the average interval of time between pulses.* Individual pulses may arrive too early or too late, but this average also is stable. The Crab Pulsar, for example, emitted pulses at an average rate of once every 0.031061537607607 seconds at midnight of January 1, 1970, and it would be doing so today were it not also steadily slowing down. Very few clocks are as accurate as the pulsars, some of which keep time correct to a thousandth of a second over ten years. Hardly anywhere else in all of physics and astronomy is there a physical quantity maintained with such regularity.

In contrast to the underlying meaning of the average pulse shape, which is still obscure, an immediate interpretation of the average rate of pulsation suggests itself: it is the rate of rotation of the neutron star itself. Only something so massive could spin with such perfect regularity. If this is true the deviations of individual pulses from this average must represent some kind of imperfect anchoring of the searchlight beam relative to the star. The beacon does not point directly away from the star. It continually shifts to and fro, sometimes pointing ahead of the rotation, sometimes behind.

Figure 17 passes on to a still higher level of complexity, and it is

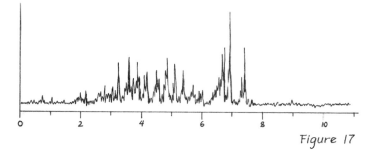

Figure 17

one that has become possible only recently through the application of exceedingly sophisticated techniques at the very largest telescopes. This figure seems to show a series of pulses. But it does not: it is a scan of a *single pulse*, and it reveals something no one had even suspected until recently. Each individual pulse is composed of a very large number of still shorter bursts of radio emission. These bursts are very brief—less than a thousandth of a second in duration—and they themselves have structure over still shorter time scales. What had appeared before as an individual pulse is now revealed to be an extraordinarily rapid sputter.

There are two ways this sputter could arise. The pulsar beacon might consist of a large number of still smaller beacons—a bundle of beacons. Alternatively it might be a single beacon which is flickering rapidly. No one knows which is correct.

So we build a picture of the pulsar searchlight beacon: violently sputtering, oscillating rapidly to and fro, waxing and waning and changing its shape, and sometimes flicking off altogether. Controlling the pattern of this beam is some process, only dimly guessed at, that maintains an average pulse shape as detailed and identifiable as a fingerprint; and controlling the rotation of the beam is the massive, heavy flywheel of the neutron star itself. What words can never convey is a full appreciation of the intensity of this beacon. Nothing on Earth even remotely approaches such an intensity. The pulsar searchlight beam is so strong that were a person to venture within a hundred million miles he would be killed—killed by mere, insubstantial radio signals—in a fraction of a second. Close to a pulsar the radio beacon is sufficiently powerful to vaporize metal and bore holes through solid rock. In a single second this beam transmits enough energy to supply the energy requirements of our entire planet—transportation, heating, industry, for Europe and America and all the rest of the world combined—for a full three hundred years.

Accompanying the radio beacon of the Crab Pulsar is a beam of visible light. Weak optical pulsations have also been detected from the Vela Pulsar, and pulsed X- and gamma-ray emissions from a few pulsars as well. But by and large they are most active in the radio region of the spectrum.

But over and above all these various forms of bursting signals, the pulsars emit another kind of radiation as well. In contrast to the bursts this radiation is continuous: a uniform, steady emission of energy in all directions away from the star. It is an exceedingly mysterious form of emission, and beyond the mere fact of its existence and the certainty that it cannot be pulsed, we know very little

about it. We do not even know if it consists of radio signals at all.
It may be cosmic rays. In fact *no one has ever succeeded in detecting
this emission*; and if it were not for one small hint, we would never
have had the slightest inkling of its existence.

The clue is the fact that pulsars are slowing down. At first sight
this might appear an entirely innocuous process, and hardly worth
paying attention to. After all, in daily life everything slows down.
Cars come to a halt once out of gas, and spinning tops soon topple.
Why get excited over the running-down of the pulsars?

The question can be put in focus by considering the steady rotation of the Earth. Clearly the Earth is not slowing down. It has been
spinning along now for more than four billion years without having
come to rest. Why does it spin so smoothly while the top does not?
Because the top is subject to friction at its tip against the ground.
So is the car, in its axle bearings and along its body against the
air. Everyday objects continually contend with frictional processes
which drain energy of motion from them and shortly bring them to
rest. But the spinning Earth does not. There is nothing in space for
it to rub against. As it is with the Earth, so it is with the other
planets, and the Sun, and so it ought to be with neutron stars. Situated in the void of interstellar space, they ought to continue rotating
indefinitely; and the fact that they are not points to some new and
unsuspected process. Somehow, isolated though they are from all
external influences, they are managing to dissipate away their energy
of rotation.

Every spinning pulsar is a flywheel, a reservoir of rotational energy;
and every one of them is steadily and smoothly converting this energy
into some other form and radiating it away into space. The intensity
of this emission is far greater than that transmitted through the
pulsar beacon itself. But no experiment has succeeded in capturing
a trace of this emitted energy: no radio telescope has detected it, no
optical observatory has ever identified its traces. No hint of its nature
has ever been found. All we know for certain is that it is there.

The concept of neutron stars had been invented more than three
decades before their discovery: given all this time to get ready, one
would have thought their properties would have been anticipated
in some detail. After all, science is supposed to be the art of prediction. But in this case the prediction failed. Prior to the discovery
of pulsars the neutron star, if it was considered at all, was considered
to be small, quiescent, difficult to detect, and for this reason dangerous to interstellar travel: something of an astronomical coral reef.

Not a single scientist anticipated their stroboscopic flashes. No one claimed they would come equipped with a handle and a bell. There is a moral in this somewhere.

But it was predicted they would be slowing down. This achievement is due to the Italian astrophysicist Franco Pacini, and his paper was published in the British journal *Nature* shortly before the discovery of pulsars. Pacini's work is particularly significant in that it laid the groundwork for all our present understanding of pulsars. Everything that has followed flows directly from it. Soon afterward his ideas were independently reinvented by the Americans James Gunn and Jeremiah Ostriker, who had been unaware of his work, and in an important series of papers they showed how many of the properties of pulsars could be understood on the basis of these ideas.

Pacini's argument begins with the recognition that *a star is a magnet*. This stellar magnetism is well known to astronomers and it is similar to the magnetism of the Earth. Figure 18 is a schematic diagram of the Earth's magnetic lines of force: they pass out through the surface of the Earth at the north magnetic pole, arc over until they lie parallel to the surface at the magnetic equator, and reenter at the south magnetic pole. At every point they run in a north-to-south direction. The needle of a compass aligns itself along these lines of force and so points north.

The stars too possess magnetic fields, and this stellar magnetism

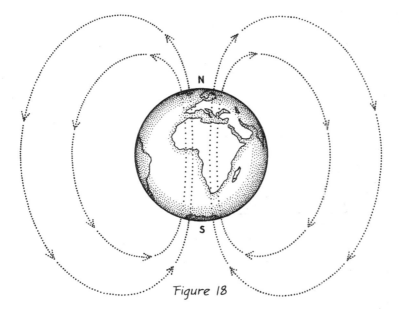

Figure 18

is responsible for many of their peculiarities. Sunspots, for example, are regions in which the Sun's field is unusually intense; and the solar flares which occasionally disrupt radio communication are the result of sudden, catastrophic readjustments in its structure.

The second step in Pacini's argument is that *if an ordinary star is a magnet, a neutron star is an intense magnet.* This is because these stars are "born" out of ordinary stars by the process of collapse. Within the body of a star the magnetic lines of force are effectively attached to the stellar matter. They cannot move across it. When the material making up the star collapses inward it carries with it these lines of force: it crushes them together and amplifies the field.

The magnetic field within an ordinary star is not particularly strong. But its amplification in the collapse is enormous, and neutron star magnetic fields are expected to be something like 1,000,000,000,000 times more intense than that of the Earth. A compass needle on the Earth experiences a mild twisting force tending to make it point north—a mere few ounces at most. The same compass on the surface of a neutron star would experience a force 1,000,000,000,000 times more intense. It would be as immovable as a boulder: rigid, fixed, locked into place with a force that no wrench or crowbar could possibly overcome.

It is this superstrong magnetic field that is the key to the understanding of pulsar radiation.

The Earth's magnetic poles do not coincide with its geographic poles. Indeed, the north magnetic pole lies only slightly beyond the coast of North America. So compass directions deviate somewhat from true geographic directions. Another consequence is that as the Earth spins it carries the magnetic poles around in a circle. The configuration is diagrammed in Figure 19, in which the angle of tilt is exaggerated for clarity. The Earth is a *tilted rotating magnet.*

The Earth is not unique in this regard. In most stars too the rotational poles do not coincide with the magnetic poles: indeed, there is a class in which the tilt angle is a full 90 degrees. Their magnetic poles lie on their geographic equators. In any case, whatever the configuration it is preserved in the collapse: neutron stars too possess tilted fields. And *a tilted rotating magnet radiates electromagnetic waves.*

The technical term for any magnet possessing north and south poles—a bar magnet, for instance, or neutron star—is a magnetic dipole ("two poles"), and the electromagnetic radiation emitted when it is spun is magnetic dipole radiation. Dipole radiation is

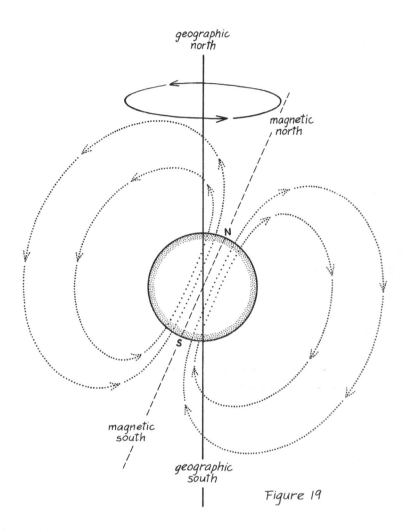

Figure 19

fundamentally a radio wave: the only difference is frequency. The familiar waves to which a radio is sensitive have a frequency of about one million cycles per second; as for magnetic dipole radiation, its frequency is just that at which the magnet is spun. If spun a million times per second, an ordinary bar magnet could be heard on a radio: a steady continuous hum, never varying, never changing in its intensity or tone. Spin the magnet more slowly and one would have to tune the receiver to lower frequencies in order to hear it. The Earth, ponderously rotating about once a day and carrying its dipole with it, is at this very moment broadcasting magnetic dipole radiation into space.

It is not hard to visualize how this radiation arises. Figure 20 returns to the magnetic field configuration and asks what would happen to the lines of force if the magnet were to be spun. Close in (within the inner dotted circle), the lines of force would follow perfectly along in this rotation, and their pattern would just be that of a rotating dipole. The farther a line of force reaches into space, however, the more rapidly it is being forced to move. Distant field lines are unable to keep up. They begin to sweep back, as within the indicated box.

The farther we progress away from the rotating magnet, the more its lines of force depart from the simple configuration they had maintained in the nonrotating case. Eventually they become swept backward into the ever-unwinding spiral pattern shown in the outskirts of Figure 20. Very far from the magnet the field configuration has lost all resemblance to that of a simple dipole: it has become an

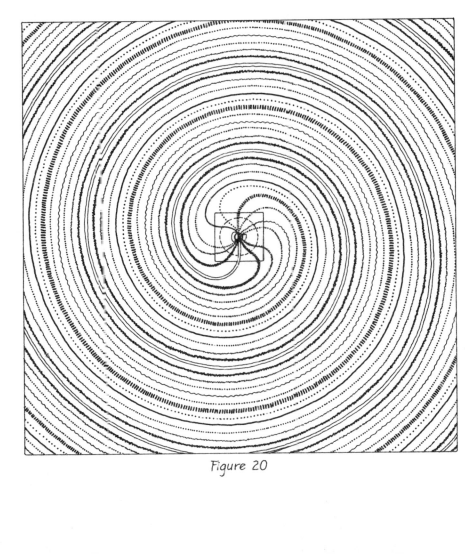

Figure 20

electromagnetic wave propagating outward at the speed of light—an ever-expanding wash of magnetic dipole radiation.

As do all other forms of electromagnetic waves, *dipole radiation carries energy*. The energy is sapped from the rotational energy of the magnet. As a neutron star spins, rapidly whipping its tilted superstrong magnetic field about, it steadily and inexorably radiates energy into space—and it slows. Not only the pulsars: every tilted rotating magnet radiates such energy, and all of them must be slowing down. But in ordinary circumstances the process is very weak. The net force acting to slow the rotation of the Earth, for example, works out to a mere 1/100,000 of an ounce. But because their magnetic fields are so intense, the pulsars are stronger radiators, and the force acting to decelerate their rotation is a gigantic 10,000,000,000,000,000 tons. So they slow.

Perhaps so . . . but perhaps not. Shortly after the discovery of pulsars, the American astrophysicists Peter Goldreich and William Julian pointed out an additional crucial feature that Pacini had failed to appreciate, and that called his work into question. At the same time it pointed the way to an entirely different explanation for pulsar deceleration. Not only that, but it opened the door to a possible understanding of the pulsar searchlight beacons.

Goldreich and Julian recognized that a *rotating magnet generates within itself an electric field*. This phenomenon is common to all rotating magnets, and it can easily be demonstrated in the laboratory. If a simple bar magnet is spun about its long axis, an electrostatic potential develops across it. If one end of a wire is now touched to one pole of the magnet and the other end to the other pole, a current will flow within this wire. The faster we spin the magnet, the stronger the current. It is a common enough process: hydroelectric power is generated in a similar way when water pressure is employed to force the magnetic rotors of hydroelectric generators into rotation.

Goldreich and Julian's point, however, was not the mere existence of this field. It was its strength. Because the pulsar magnetic field is so strong, it generates a correspondingly enormous electric field. They showed that this electric field, in contrast to those generated under more common circumstances, is so powerful as to tear charged particles out of the surface of the neutron star and propel them into space.

Electric fields exert forces on charged particles. In any piece of matter, whether it be a strand of copper in a wire or the surface of a neutron star, two and only two types of charge are present: the positively charged nuclei of atoms and the negatively charged electrons orbiting about these nuclei. Depending on the circumstances, these particles may or may not be able to resist this force. For example, within the strand of copper making up ordinary electrical wiring, the positively charged nuclei are rigidly held in place, and only the electrons are free to move. Their motion is the electric current.

These particles are experiencing a relatively mild force—they stay in the wire. But on a neutron star they do not. The material making up the star is totally powerless to retain them. It shreds. Under the action of the superstrong pulsar electric field, both positive and negative charges boil from the surface of the star, and they are violently accelerated away into space. Within an inch in their motion they are boosted to nearly the velocity of light. They fill the region around the star in great numbers. The pulsar has created its own atmosphere. But this is no ordinary atmosphere. It is a region of space permeated with powerful electric currents, intense radiation levels, and atoms torn into their constituent parts. The technical term for such a gas is a plasma; and this plasma surrounding the star, embedded in its magnetic field, constitutes not the *atmosphere* but the *magnetosphere* of the neutron star.

The structure of this magnetosphere is so complex that no one has succeeded in unraveling it. In particular its effect on magnetic dipole radiation is not known. The simple picture of a rotating magnetic field Pacini employed cannot be correct if the field coexists with a plasma. The plasma alters the field. We do not understand the physics well enough to decide whether the new configuration is even capable of emitting dipole radiation when spun. The Goldreich-Julian charges may conceivably short circuit the emitter and render Pacini's mechanism inoperative.

Whether or not they do, the pulsar magnetosphere has another important property as well. It contains a good deal of energy. This energy is that of motion of the charges out of which it is made, and it is very large since they are moving so rapidly. As always in such a circumstance, we must ask just where this energy has come from. Finding a region of space chock full of energy is like finding a room chock full of dollar bills. We are anxious to learn how they were obtained.

Goldreich and Julian succeeded in showing that the magnetospheric energy is sapped from that of rotation of the star. The steady streaming of particles away from the surface produces a reaction back against the star which acts to slow its spin. Even if rotating magnets do not emit dipole radiation, the Goldreich-Julian process in and of itself will account for the deceleration of the pulsars.

One fact with two possible explanations: whether Pacini's or the

Goldreich-Julian mechanism is closer to the truth could be decided by observation—by capturing a trace of the emitted energy. Until this is done, or until some clever theoretician succeeds in unraveling the magnetosphere in all its complexity, the question will remain in limbo.

We come at last to that aspect of pulsar emission which is at once the most obvious thing about pulsars and the most poorly understood: their searchlight beacons. They must arise somewhere within the magnetosphere. How?

They must be something completely different from the emission responsible for the slowing down of pulsars. This is obvious in the case of the Goldreich-Julian mechanism, which consists of particles, not waves. It is equally true of the Pacini mechanism, for the magnetic dipole radiation is emitted at a frequency far lower than that at which pulsar emissions are observed—one cycle per second for a one-second pulsar, for example, as opposed to millions of cycles per second for the waves making up its radio bursts. Furthermore, magnetic dipole radiation is emitted continuously and in all directions at once.

We will have to look elsewhere.

What makes radio waves? Rotating magnets do, but this is no help here. Also, it is a relatively uncommon process in the natural world. Far more prevalent is the generation of radio signals by electric charges.

A charged particle at rest does not emit. The same is true if it moves with constant velocity in a straight line. It makes no difference how rapidly the charge may be moving, ten miles per hour or ripping along at just under the speed of light: so long as the state of motion does not alter, the charge emits no signals. But if the particle *accelerates* in any way—if it collides with another, bringing itself suddenly to rest; or if it bounces off a wall, preserving unchanged its speed but altering its direction; if some outside agency acts to increase its velocity or bends its path in a circle—if any of these things occurs, the charge radiates electromagnetic waves. This is how radio and television stations broadcast their signals. Within their antennas electric currents are accelerated alternately one way and then another: continually changing their velocity, the electrons making up these currents continually emit.

In what direction is the emission? The answer depends on the charge's velocity. For low velocities it emits in essentially all directions. Electrons in broadcasting antennas travel relatively slowly, and this is what they do. These stations broadcast in almost all directions at once. But the charged particles making up the pulsar magnetosphere move at very nearly the speed of light, and for them the answer is different. They broadcast in a narrow beam pointing straight ahead—in their direction of motion.

In what direction do the charges making up the pulsar magnetosphere move? They are guided in their motion by magnetic lines of force. Charges in a plasma can never move across these lines: they can only move along them. The charge is a bead; the magnetic line of force is a wire. The bead slides along the wire.

These are the physical principles from which any theory of the pulsations must be built. There are some tantalizing resonances. We need charged particles: the magnetosphere is full of them. We need a beacon: each such charge is capable of emitting one. Now we need acceleration, a change in their state of motion; the more the better.

Figure 21 reproduces Figure 19 illustrating the rotating pulsar

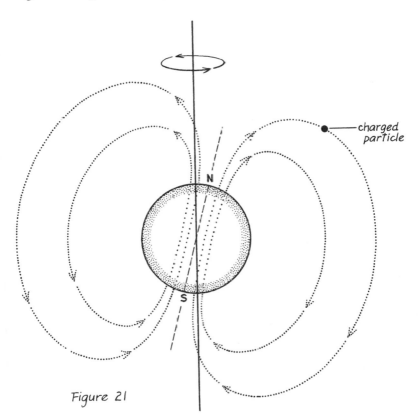

Figure 21

magnetic field but with one addition: that of a single charged particle. There are many such charges making up the magnetosphere but for now concentrate on it. Is it moving?

The magnetic field lines are certainly moving. They are *rotating,* precisely in synchronism with the star. As for the charge, it cannot move across these lines: it must be rotating too, swinging about the star, a bead on the spoke of a spinning wheel.

The star is rotating uniformly. Is the charge moving uniformly? Or is it accelerating? Follow its path. At the instant shown in Figure 21, it is moving into the page. Half a rotation later it has swung around to the other side of the diagram. Now it is moving out of the page—it has reversed direction. The velocity was not constant. When did the reversal occur? At no particular instant: the charge reversed smoothly, steadily and continually arcing about. At *no* point in its motion was the charge's velocity steady. It radiated electromagnetic waves continuously.

This radiation is in a beam, and at the instant shown in Figure 21 the beam points into the page. As the star spins about, it does too. Half a rotation later it points out of the page and we are illuminated by its emission. *At this instant, as the beacon sweeps by, we receive a pulse.*

The model nicely accounts for the existence of pulsar bursts. We can also understand their great regularity in time, for within it the rotation of the beam is rigidly tied to that of the star. But how to account for the deviations from regularity from one pulse to the next exhibited by the observations?

The beam swings steadily only if the motion of the charge is precisely in a circle. But is it? The motion is circular if the charge keeps the same distance from the star. Suppose it does not. Which way does the beacon point then? Imagine that the charge were to be traveling outward along its magnetic field line at an enormous velocity—far more rapidly than its velocity imparted by rotation. In this case the motion would be essentially outward away from the star. The beacon would point not into the page, but along the page: up and off to the right in Figure 21. In the more realistic case of a slow outward motion, the beacon would deviate only slightly to the right of a line pointing directly into the page. Motion of the charge inward toward the star would swing the beacon to the left of into the page. The pulse-to-pulse irregularity in timing of pulsar bursts can be understood by imagining that the charge continually jitters back and forth along its field line, swinging its beacon slightly to and fro about its equilibrium configuration. Why does it do this? We have no idea. The observations are telling us that it does.

The model looks good—it reproduces a number of the features of pulsar emission. But it suffers from a flaw. The intensity of radiation from the charge falls far short of actual pulsar intensities. The predicted signal strength is too weak—by enormous factors. The model fails.

How to save it? We need to boost the emission rate. This means we must boost the charge's acceleration, for the stronger this acceleration, the more intense the emitted signal. We can do this by imagining the charge to be located farther and farther from the star. The greater the charge's altitude above the star, the greater its acceleration imparted by rotation.

But there is a limit to how distant it could be. At some critical distance from the star the rotating magnetic field must be forcing it to travel just at the speed of light. Beyond this point something has to break down, for rigid corotation would lead to faster-than-light velocities, which are impossible. Much of present research on the pulsar magnetosphere concentrates on its behavior at this critical *speed-of-light distance*, and it aims to understand just how the rigid rotation of the inner zone breaks down . . . and what it breaks down into. But at any rate there is clearly a maximum intensity of the pulsar beacon allowed for by the model, and the problem is that it lies far below what the observations reveal.

The picture cannot be rescued in this way. Try another way. Many charges acting together will emit more strongly than one. The emission intensity turns out to depend on the square of the number of particles: a bunch consisting of two charges emits four times as strongly as a single charge; one hundred charges, ten thousand times as strongly. This method works. We save the picture by postulating some kind of *thing*—a coherent unit, a collection of many charged particles acting together, and orbiting far from the neutron star at close to the speed-of-light distance.

This object must be moderately large, and contain a net charge sufficient to boost the predicted emission intensity up to observed values. On the other hand, it cannot be solid: the pulsar magnetosphere is sufficiently hot to vaporize anything within. It must be gaseous—a cloud. However, unlike ordinary clouds it must maintain itself as a permanent unit. Quite possibly it has a structure, a shape.

Could this shape be related to that of the received pulse? A large, diffuse structure will emit a broad, diffuse beam, and we will register a broad pulse as it sweeps by. More compact structures yield sharper bursts, and structures with complex shapes complex bursts. The fact that individual pulses differ greatly from one to the other means that the bunch must be fluctuating, continually and rapidly changing in structure. The fact that *average* pulse shapes are so steady in time means that some mechanism must be at work preserving a similar average structure toward which the bunch constantly tends, but away from which it is constantly forced. And this average structure must differ from one pulsar to another.

As for what determines this average, it may be significant that the pattern of the Sun's magnetic lines of force differs radically from that of the Earth. It is far more complex. Presumably the same is true of those of other stars as well—neutron stars included. Because charges are guided in their motion by these patterns, it is reasonable to believe that the structure of the bunch is dictated in some way by its details. The pulsar magnetic field might be mapped, if only we knew how, through its average pulse shape.

The charged-particle bunch is a mysterious object: gaseous, complex, fluctuating incessantly about some equilibrium form, whipping about the pulsar at nearly the speed of light. No one would have invented it had we not been forced to in an effort to rescue the model. But this is far from saying the bunch exists.

Perhaps the model cannot be rescued.

To put things into perspective, recall the analogy between the pulsar magnetosphere and the atmosphere of the Earth. It is, in fact, not so bad an analogy at all. The magnetosphere is a gas surrounding the neutron star; air is a gas surrounding the Earth. And air never clusters together in this way. After all, if it did we would find ourselves passing from regions of near-vacuum into clumps of excessively high pressure every time we went out to take a stroll. Quite the contrary: air spreads itself out uniformly, and it actively resists any attempts to compress it together into bunches.

Why should the pulsar magnetosphere be any different?

A further reason for doubting the existence of these bunches is that they must be electrically charged, and like charges repel one another. All in all, there are a number of difficulties raised by postulating such structures. Those who believe—and this is the appropriate term—in the picture we have been drawing are faced with the task of finding some mechanism whereby the bunches are formed and maintained against the opposing forces of pressure and electrical repulsion. They have a difficult task before them, and it is not at all clear that they will ever be successful. So far they have not. At the time of this writing no one has been able to show that such a mechanism is even possible. And until such a demonstration is made, it remains possible that everything we have been saying is wrong.

It may be we have been barking up the wrong tree altogether.

*   *   *

Try another model.

The pulsar magnetosphere is full of charges. They are moving at nearly the velocity of light. Where did they come from? They came from the surface of the star—and while on this surface they were at rest. These charges must have been accelerated somewhere, and during this period of acceleration they must have radiated electromagnetic waves. Could this be the pulsar emission?

We will only be able to make this model work if we can succeed in showing that the emission must be beamed. Is it? The laws of physics hold that the radiation is emitted in the direction of motion of the particles. Is there any reason to believe that when these particles are accelerated away from the surface of the star, they all move in the same direction?

The laws of physics also hold that when these charged particles move, they move along the magnetic lines of force. They move like beads on a wire. Take another look now at the tilted, rotating magnetic field structure illustrated in Figure 21. *There are only two regions in which the lines of force point directly away from the star: the north and south magnetic poles.*

Consider two charged particles lying on the surface of the neutron star. Choose them carefully: the first will lie at the north magnetic pole, the second somewhere along the magnetic equator. An electric field acts to lift both of them away from the star. The charge at the magnetic pole is free to move, and it accelerates vertically upward along a line of force. But the second charge is not free to move outward. If it were to do so, it would cross a magnetic field line, and this is forbidden. The motion of the charges is beamed. So is the radio emission.

A more careful analysis shows that the acceleration is strong only for particles located quite close to one or the other of the two magnetic poles. So the picture naturally accounts for the narrowness of the pulsar beacon. Here too one is forced to postulate some bunching mechanism to account for observed pulsar intensities, but because the vertical acceleration is so violent many workers are of the opinion that the task is easier in this case. The picture might be made to work.

As before, we must now attempt to account for each and every feature of pulsar emission within the framework of this new model. What additional features must be built into it in order to understand the rich complexity of the pulsations that is actually observed? For example, what is the significance of the average pulse shape within this picture? Does it represent in some way a pattern etched upon the very surface of the neutron star itself? Experiments have shown that charged particles are more easily emitted from the tips of needles than from flat surfaces. Is it possible that charged particles are similarly emitted from high points of the neutron star terrain? Do pulsars with simple average pulse shapes possess simple hillocks at their magnetic poles, and those with more complicated average shapes more chaotic terrain there? Alternatively, is it more likely that the average pulse profile is determined by the detailed pattern of the pulsar magnetic field structure?

The pulse-to-pulse fluctuations in beam shape might represent a continually fluctuating pattern of particle emission from the surface of the star. Some workers have proposed that the magnetosphere might react back upon the star that made it. They envisage sudden discharges onto the star—lightning bolts. Could each bolt knock out a new spray of particles, and so trigger a new beam pattern? Does the erratic fluctuation in the timing of pulses—some early, some late—represent a rapid swinging to and fro of the beacon produced by successive bunches knocked loose in slightly different directions? Or do the field lines themselves wave back and forth like strands of seaweed?

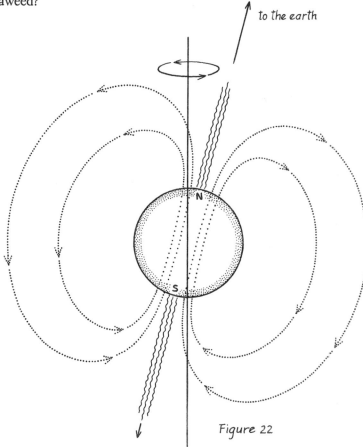

to the earth

Figure 22

The pattern of radiation emitted by this model is illustrated in Figure 22. A striking feature is that it predicts *two* beams from every neutron star: one from each magnetic pole. Each beacon sweeps out a cone in space as the neutron star spins, but only one beam intercepts the Earth in its sweep.

But certain pulsars exhibit interpulses—subsidiary bursts located between the main bursts. It must be significant that the interpulses usually occur about halfway through the pulsar cycle. The configuration shown in Figure 23, in which the magnetic poles lie on the geographic equator, may offer an explanation for this phenomenon.

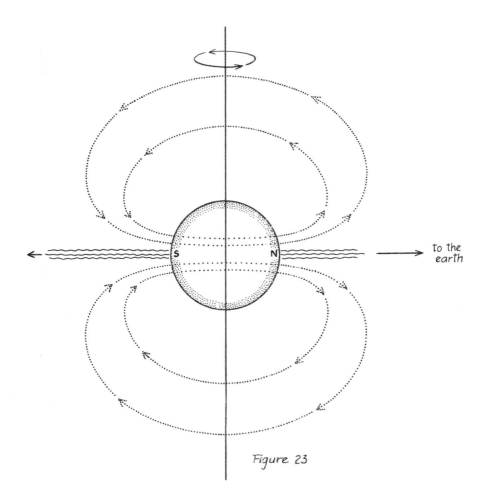

to the earth

Figure 23

It is a puzzle—a game. The pulsar astronomer tries a model. He tries another. He fiddles with the picture, modifies it, adjusts it to fit the observations. He is nothing if not inventive. In other circumstances what he does might even be called cheating: cooking up an explanation, or several explanations, and all after the facts are in. Scientists are supposed to be more rigorous, working logically from the fundamental laws of nature to an inevitable conclusion. The kind of fiddling that actually goes on seems unprofessionally sloppy at times.

So be it. The pulsar astronomer sits in the office, feet upon the desktop, staring out the window. On the desk unfinished business lies untended. He is woolgathering. Idly across his mind a technical paper he read last week floats by—some new observation of pulsar properties. It reminds him of something a colleague had mentioned the other day, something about plasmas, magnetic fields. . . .

Pulsar models are invented over lunch and discarded by quitting time. Some survive longer, a few long enough to make it into print. It is remarkable how much of the effort is conducted in conversations. The process often seems to go best as a dialogue. The phone bills mount.

The pulsar astronomer flies off to a conference. Lectures are scheduled. Participants present the latest data and argue rival theories. He listens dutifully—sometimes. But the action in the hallway beckons too. A group has gathered around the coffee machine and one member is pushing for the speed-of-light-distance model. Everyone else is jumping on him. Someone says something that sticks in his mind . . . he had never thought of that before.

It has been going on this way for more than a decade now.

We are left in an unsatisfactory state. So far as pulsar deceleration is concerned, there are two theories and no observations. As for the pulsar beacon, there are two theories, each with innumerable variants, and plenty of observations—some would say too many. Somewhere in all that mass of data must lie the key, the clue that will lead to the truth. Somewhere amid the theories must lie the final unifying insight bringing order to the wealth of observational material.

Meanwhile, oblivious to our ignorance, the pulsars twinkle on. Rotating wildly in the void of space—one rotation a second, thirty rotations a second—they surround themselves with a superheated plasma atmosphere. Out of this inferno a powerful beam of radio signals is generated, and it spins about in synchronism with the star. Thousands of light years away, thousands of years later, faint traces of these beams flick across the face of the Earth. They are coming to us out of the blaze and sparkle of the electromagnetic storm.

5

Glitch

In the winter of 1969 the Vela Pulsar underwent an extraordinary transformation, the likes of which had never before been observed in any astronomical body. It took the scientific world by storm. It also took *me* by storm, and in this and the next chapter I want to describe the efforts to understand this event from a personal perspective. From the beginning my involvement with pulsars has been colored by this strange event, and I am still not free of it.

On February 24, 1969, two radio astronomers in Australia observed the Vela Pulsar. They found it to be in an entirely normal state. By chance Vela was not observed again for a full week. When it was next monitored—from an American observatory this time—it was also found to be in a normal state, but with one important difference. It was pulsing more rapidly. At some point within that seven-day period the Vela Pulsar had reversed its steady process of slowing down and had sped up.

Nothing in the emission prior to the event had given any indication of what was about to occur. The transformation seemed to have occurred without the slightest warning. But it did leave its subsequent mark upon the pulsar. After Vela had sped up it resumed its steady process of slowing down—but more rapidly than before. Its rate of deceleration had increased.

Slowly, smoothly, this rate of slowing down changed. The Vela Pulsar began to slow more slowly. Two months passed. Six months passed, and then a year. Every month the rate of slowing down was less. Every month the rate of slowing down approached a little more closely the value it had had long ago, before the pulsar had sped up in the first place.

The faraway planet of X—— orbits about a star quite similar to our own, but it lies much closer to it. As a consequence the planet is very hot. Monstrously huge, filling the sky, its sun glares down and has baked X—— to a temperature that we would find unbearable. The temperature of this world, in fact, actually exceeds the boiling point of water at every point upon its surface. If a glass of water were to be set down on the ground in daylight, it would immediately begin to boil, so intense is the blaze of sunlight. The same glass set outside in the middle of the night would also boil in the presence of the planet's superheated atmosphere. It would boil if placed upon the north pole in the middle of the perpetual night of winter, or on the summit of the highest mountain.

Planet X—— is therefore dry. It is not just somewhat dry, like the Sahara Desert or the floor of Death Valley. It is absolutely dry. Nowhere upon it is to be found even the smallest of ponds, the most meager of brooks, or the barest hint of an oasis. The surface of X—— is a wasteland of barren plains and blowing sands, and of harsh and desiccated mountains. It is an abode that would seem utterly inimical to life.

And yet, strange to say, life does exist on this world, and not just life but intelligent life. Of course, the inhabitants of X—— do not look much like us, with their multiple heads and their thick, scaly hides. Their bodies contain no liquids of any sort, and their very biochemistry is alien. Nevertheless, they are not so very different from us as all that, and many of their activities are things that we can appreciate. Like us, they have a lively nature and enjoy nothing so much as a good party from time to time. Like us, they tend to congregate together in cities and spend far too much of their time stuck in traffic jams. And like us, some of them are scientists.

Of these scientists, some are chemists, and they have conducted experiments on molecular compounds. They have taken the elements hydrogen and oxygen and, in the laboratory, combined them to form molecules of H_2O. They have studied this molecule's properties. They have taken its spectrum. They have determined the energy required to dissociate it. They have learned that it is a polar molecule. They have done more. They have synthesized these molecules in large numbers and injected them into a chamber, and they have succeeded in cooling the chamber so that tiny puddles of liquid have collected on the floor. In this way they have learned that liquid H_2O is transparent and colorless, that it flows easily, and that it has a shiny, mirrorlike surface. They have measured the density, the viscosity, and the surface tension of this liquid. The experiments, of course, had not been easy, for the molecules were difficult to synthesize, and it was harder still to refrigerate the test chamber against their world's blaze of heat. At the very most a mere few ounces of liquid water were made available for study.

Finally, in a supreme effort, the scientists of planet X—— succeeded in refrigerating their chamber so intensely that the water within it briefly froze. In this way they learned that water expands slightly upon solidifying; that solid water is brittle, easily fractured, and smooth.

And now we ask a question about these hypothetical beings: is there any way they would be able to imagine our own planet Earth?

Is there any way that without ever having seen our world they would be able to predict its properties from what they have learned in the laboratory? Has their scientific knowledge led them to a realization that water sometimes flows in gentle babbling brooks, and on other occasions in mighty torrents? Has anything they have learned prepared them for the infinite moods of the ocean: placid and blue one day; ominous, heaving, and gray on the next? Could they predict from the laws of nature they have uncovered the sparkle of sunlight on a pond? Could they predict that no two snowflakes are exactly alike; that snow is shining white when it first falls in the city, but dark and dirty soon after; that it crunches underfoot; that it is a pleasure to ski on but a nuisance to traffic? Could they predict that most of our major cities are built on coastlines or rivers? And what would they say when presented with an umbrella or a surfboard?

All of these properties of solid and liquid water can *in principle* be predicted from a knowledge of the properties of the H_2O molecule together with observations performed on a small thimbleful of water. But to say something is possible in principle is not to say it is easy to do. It is not even to say it will ever be done. It is clear enough in practice that the scientists of planet X—— would have a very hard time of things indeed; and it is likely that they would succeed in predicting some things about our world, fail in predicting others, and in many cases manage to predict things that are purely and simply wrong.

We are in the same situation with regard to neutron stars. In one sense it might be thought that a neutron star is not so very alien an object as all that, for it is composed of the same elementary particles as an atomic nucleus. The atomic nucleus in fact is no more than a microscopic bit of neutron star matter, and we do, after all, understand much about it. The only real difference is size. But although all this is perfectly true, it misses the point. The point is that large quantities of any substance behave radically differently than small quantities do.

We have available to us for study in the laboratory the individual elementary particles out of which neutron stars are built: neutrons, protons, and electrons. We also have available for study combinations of relatively small numbers of these particles: the nuclei of atoms. The largest such nucleus known is that of the element Mendelevium: it contains 155 neutrons and 101 protons. It is very large as atomic nuclei go, but compared to a neutron star it is infinitesimal. We study its properties We study the properties of the lighter nuclei, and those of the isolated elementary particles. And then, using the body of knowledge that we have acquired as a springboard, we make a leap, an enormous risky jump, and we try to build a picture of a neutron star.

How do we do this? We do it in the same way that the inhabitants of the hypothetical planet X—— would construct a picture of the Earth. It is not only a matter of solving equations or performing experiments. It is not only a matter of finding answers to questions. More than anything else, it is a matter of asking the right questions. It is a matter of guessing which factors might be important and which factors might not. And above all, we are required to perform an act of imagination. We must create in our minds a world that is utterly alien utterly unfamiliar. That is easy enough to do in science fiction. It is another matter entirely if you have to be right.

All this had been known by scientists in an abstract sort of way since the 1930's, when neutron stars were first proposed. But it is one thing to know something intellectually and another to feel it in your gut. It was not until the discovery of pulsars in 1967 that scientists began to realize just how bizarre a thing this so-called giant nucleus of theirs could be. And it is safe to say that it was not until the winter of 1969, when the Vela Pulsar sped up, that the full magnitude of the realization came to be felt.

The event was not large. The rotation rate of the pulsar increased by a mere two parts per million. If this were to happen to the Earth, the day would be only 0.2 of a second shorter. But this does not happen to the Earth—ever. It does not happen to the Sun, or to any other astronomical body. It was particularly extraordinary in light of the truly remarkable regularity of the other pulsars, and of the Vela Pulsar itself up until the moment of the event. It was immediately and widely appreciated that what had appeared to us as a sudden increase in speed of rotation was actually a relatively unimportant consequence of some cataclysm that had occurred within the star— a cataclysm new to astrophysics and the nature of which was entirely unknown. This cataclysm was one which apparently never occurred in any of the more familiar objects of the universe, and it clearly arose from some property of neutron stars common to them and them alone. It was this more than any other discovery that sparked a widespread interest in the study of the internal constitution of neutron stars, and it launched the community of physicists and astronomers on the effort to understand these strange new worlds.

This enterprise has been led by the Columbia University physicist Malvin Ruderman. More than any other single person, it is he who is responsible for our present understanding of pulsars and neutron stars. Not that everything we know about neutron stars has been discovered by Ruderman; the field is too big and involves far too many people for it to be monopolized by any single person. Nor has Ruderman even been right in everything he has proposed. Good science does not always involve being right—at least not in every detail. It is far more important to paint the broad outlines of a subject, to discover entirely new phenomena, and to point the way to new and fruitful investigations. In this Ruderman has been pre-eminent. Science is too much of a democracy and scientists too argumentative and fiercely independent for anyone to direct its progress in the ordinary sense of the word. Influential scientists such as Ruderman do not lead by giving orders. Rather, they lead by inspiration. They lead by the power and the authority of their ideas. Time and again it has been Ruderman who has pointed to the essence of a phenomenon; who has done the new, the surprising, the imaginative work. He has kept us on our toes.

The picture that has slowly been evolved, by Ruderman and by others, of the internal structure of neutron stars is of one of the strangest objects known to science. Nowhere else in the universe is there anything remotely like it. Somewhere within this picture must lie the clue to the Vela Pulsar's strange behavior.

The average density of a neutron star is about that of the atomic nucleus. But the star is not this dense everywhere. If one were to start at its surface and tunnel in toward the center, one would encounter matter at progressively greater and greater densities. The terrible weight of the overlying layers of the star crushes the matter in its deep interior. If matter is progressively crushed in this way, there are a number of points at which it undergoes profound changes. These critical *densities* are analogous to the two critical *temperatures* of water: that at which it freezes, and that at which it boils. A neutron star has a somewhat layered, onionlike structure, and the deeper one penetrates into it, the stranger it becomes.

To visualize most easily the nature of the matter within neutron stars, it is convenient to conduct an experiment in our minds. Begin with a chunk of ordinary matter—rock, say—and progressively crush it to higher densities. As we do so it will make a series of transitions to progressively stranger and stranger states, mimicking at each stage the matter at progressively deeper points within the star.

Begin with a cube of rock one mile on a side. Then move an array of giant battering rams against this cube, and compress it until it is 400 feet high. The cube is now more dense than any earthly material. It would be all we could do to pick up a piece of it a mere one inch on a side, for this cubic inch weighs nearly 400 pounds.

Next recall that, unlike ordinary stars and planets, neutron stars possess superstrong magnetic fields. To duplicate conditions within the star, apply such an intense field to the cube. This magnetic field is so strong that it distorts the very atoms out of which matter is made. In the absence of a magnetic field, atoms are spherical, but in superstrong fields they become pencil-shaped. These "pencils" arrange themselves to point along the magnetic field lines, lined up like so many needles laid end to end. They exert chemical forces on each other and bond into long, thin molecular chains. The matter has assumed a stringy, hairy structure. This is *the first critical stage of compression—the surface of the neutron star.*

The cube, initially one mile high, has now been crushed down to 400 feet. Crush it further, until it is a mere 20 feet high. Now each cubic inch of this superdense stuff weighs 2,000 tons, and in this process it has made a transition to an entirely unfamiliar state.

The atoms of which ordinary matter is made have been crushed out of existence. They have been forced to overlap. Atoms, whether spherical or needle-shaped, are composed of electrons orbiting about nuclei, but once they have been crushed together, this orderly structure is destroyed. It is like what would happen if two brick buildings were to be forced together. They would decompose into their constituent parts—into the bricks. This is *the second critical stage of compression,* and in this stage matter is dissolved into a uniform, homogeneous mixture of the components of atoms: electrons and nuclei. The matter is entirely without chemistry. It cannot burn, for example, and it is neither acidic nor basic, and it does not have a taste. All these are purely chemical properties of matter, and chemistry arises from the interactions of atoms—but the atoms are gone.

This matter forms a solid. It does so for reasons that involve the forces which nuclei exert upon each other. These forces are very simple. Nuclei possess a positive electric charge—and like charges repel one another. So the nuclei try to avoid each other. The most favorable state is one in which each nucleus is situated at the greatest possible distance from all its neighbors. The assemblage, with each particle repelling and being repelled by all the others, acts for all the world like a crowd of people jammed together in a subway. In their effort to avoid one another, they *hold still*. Each nucleus finds the location which places it farthest from its neighboring nuclei, and it stays there. The matter has frozen: not because it is cold, but because it is dense. Neutron stars, like the Earth, possess an *outer crust*. This crust commences a mere few yards beneath the surface of the star, and it extends downward a matter of miles.

The cube, initially one mile on a side, is now 20 feet across. Compress it still further. As we do so, the nuclei start to absorb electrons. An atomic nucleus consists of roughly equal numbers of neutrons and protons: the protons now react with the absorbed electrons to form still more neutrons, forced to do so by the compression. Slowly and smoothly, ordinary matter is being squeezed into neutronic matter.

Compress the cube until it is two feet on a side. Then each cubic inch weighs two million tons. It is still a solid, and by now it consists almost entirely of neutron-rich nuclei, together with a few residual electrons. But at this density we encounter the *third critical stage* of compression and the neutrons begin to boil out of the nuclei. The nuclei have become so rich in neutrons that they find themselves unable to contain them all; and one by one at first, but then at ever-increasing rates as the density is increased still more, the neutrons escape from their nuclei like so many bees from a hive. They fill the spaces between the nuclei. They move freely about. They flow. They form a fluid—a *superfluid*.

The crust of a neutron star, though unusual, is at least solid, and solids are things one encounters in daily life. But nothing in daily experience exhibits the properties of a superfluid. Indeed, upon the Earth there is one and only one superfluid known, and it is very rare. If ordinary helium—helium from a balloon—is refrigerated to a temperature of four degrees above absolute zero, it liquefies. This transition is precisely analogous to the transition water vapor undergoes when it is cooled below 212 degrees Fahrenheit, and there is nothing particularly remarkable about the resulting liquid helium. But if this liquid is then cooled still further, to two degrees above absolute zero, it makes another kind of transition and turns from an ordinary fluid into a superfluid.

The most striking property of superfluid helium is a complete absence of viscosity, the property whereby swirling motions in fluids are forced to die out. Water has a moderate viscosity, and if we were to give a bathtub full of water a stir the motion would persist for a matter of minutes. Honey has a large viscosity and swirling motions in it die out immediately. Superfluid helium, on the other hand, has *no* viscosity, and if we were to stir a bathtub full of superfluid the resulting motions would quite literally continue for months. Stir it in summer and return in the fall: it would still be churning.

Beyond the third critical stage of compression, matter consists of a solid *and* a superfluid, the two existing together. The neutron superfluid interpenetrates and flows through the solid. We are describing here the *inner crust of the neutron star*. It lies just beneath the outer crust, and through it moves the neutron superfluid—an underground ocean.

Still more compression. Crush the cube until it is two inches on a side. Into this space four billion tons of matter have been forced. The nuclei are now so close together that they touch. They merge. They blend together and lose their identity. Beyond this *fourth critical stage of compression* the nuclei have entirely disappeared into a uniform soup consisting almost entirely of a neutron superfluid, together with trace quantities of free electrons and protons. The solid has been dissolved by the compression. At this stage we have reached a point lying perhaps halfway from the surface to the center of the star, and this point marks the inner boundary of the crust. Below this boundary, and stretching deep down into the interior, lies an ocean of superfluid neutrons.

Swim through this ocean still deeper into the heart of the star. The density does not, in fact, increase by all that much. In terms of the hypothetical experiment the conditions at the very center of the star correspond to shrinking the cube to one-quarter its present size. It is a relatively modest increase in density. But as a result of this increase an important thing happens.

We run out of understanding.

With this increase in density a multitude of elementary particles appear within the star. The denser the star, the more rapidly do the neutrons within it move; at its center they are moving so rapidly that a spray of new particles is produced every time they collide. On the Earth these strange particles are created only rarely, in experiments in giant particle accelerators. But in the star it happens all the time.

Elementary particle physics is a field lying at the very frontier of knowledge today. Literally hundreds of exotic particles are known; none is understood in any detail. The reason is they do not persist long enough to be properly studied. Like fireflies, they are evanescent.

Once created in an accelerator, they decay—into other exotic particles, which themselves survive only briefly before decaying in turn. The pi meson, for example, survives a mere 300 millionths of a second on the average, and it is long-lived as such things go. Nevertheless, in their brief existence they exert forces of great complexity upon each other, and they interact in diverse ways.

These new elementary particles decay in the laboratory—but not in a neutron star. At great densities they become stable. They flood the deep interior in enormous numbers. The very center of a neutron star is composed of matter whose properties we understand hardly at all.

But there is more. *The matter is more dense than an elementary particle.* The compression has become so great that the fundamental units out of which matter is made are crushed together. Everything in daily experience, even something as dense as a block of lead, contains a good deal of empty space within it. The individual particles out of which ordinary matter is made do not touch. The same is true of the heart of the Sun, or deep within the planets. But in a neutron star, matter is totally packed: no more empty spaces. But even at this stage we have not yet reached the center of the star. At deeper levels the elementary particles are pressed even closer together. . . .

There is nothing new about this situation. It first occurred just below the star's surface, and there it was the atoms that were crushed together. A bit further down, at the base of the crust, it was the nuclei that were forced to merge. In both cases the structure dissolved into its constituent parts. But what does an elementary particle dissolve into? Does it *have* any constituent parts?

The terrible pressure at the center of a neutron star has forced upon us a question. This question lies at the heart of modern physics and it is unsolved at present. It is the question of the ultimate constitution of matter. Can it really be true that matter is composed of hundreds of different types of elementary particles? Or are these particles themselves built out of still more fundamental units?

The weight of opinion today is that all so-called elementary particles are themselves composed of quarks. If so, then at its very center a neutron star is not made out of neutrons at all: it is made out of quarks. These quarks, in turn, are exceedingly elusive beasts. Not a single one has ever been directly captured and studied in the laboratory. In spite of the most intense efforts, this particle, supposedly the fundamental building-block of matter, has remained an enigma.

Many suggestions have been put forward concerning the nature of matter at the heart of a neutron star. It has been proposed that this matter solidifies—that neutron stars possess solid cores, as well as solid crusts. It has been proposed that great numbers of charged pi mesons are present, carrying superconducting electrical currents without resistance. It has been proposed that matter makes a transition to what is blithely described as an "abnormal state," in which elementary particles act as if they had no mass. But every one of these proposals is hedged with uncertainty.

No one knows.

So we complete our journey from the surface to the center of a neutron star. It was a journey of stages, where from time to time we paused to marvel as matter underwent a series of transitions. Each of these transitions seems more extraordinary than the one before, and each leaves the material almost unrecognizably different from the plain and ordinary stuff with which we began: a cube of rock one mile on a side and now, at the completion of the experiment, compressed to such frightful densities that the same quantity of matter occupies a region of space smaller than one's thumb.

In Figure 24 the picture we have arrived at is summarized. First, the surface of the star, stringlike in its microscopic structure and

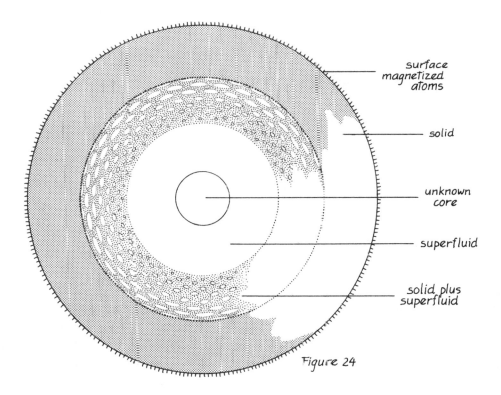

surface
magnetized
atoms

solid

unknown
core

superfluid

solid plus
superfluid

Figure 24

composed of needle-thin atoms. Beneath this surface the crust, incomparably stronger than steel. Somewhere deep within the crust an underground ocean appears, an ocean of superfluid. Far beneath this point the crust dissolves and below stretches the vastness of the superfluid. And finally, at the very heart of the star, a core of largely unknown properties. Nowhere in the universe is there to be found a piece of architecture remotely approaching this picture in its strangeness and unfamiliarity. Such is the effect of compression.

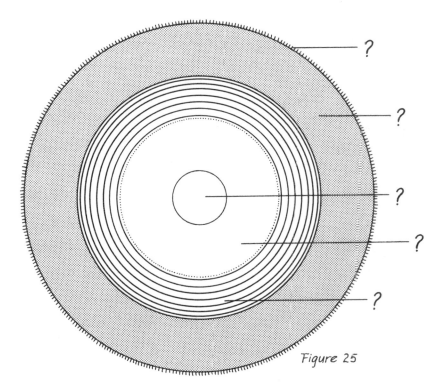

Figure 25

Finally, Figure 25 presents a somewhat different picture of neutron star structure. Perhaps it is the more accurate of the two.

In the spring of 1968, when the discovery of pulsars was announced in the British journal *Nature*, I was enrolled as a graduate student in physics at Yale, and I was engaged in putting the final touches on my Ph.D. dissertation. Like many, I still remember my first sight of this now famous issue, with the words "Possible Neutron Star" on its cover. I also recall that my response to this news was a shrug.

There were a number of reasons for my unenthusiastic reaction. First, there was the matter of my position at the time: I was a beginning scientist who had just completed the research required for his degree. Students who have successfully completed their first piece of original research are often inordinately proud of it, and tend to regard it as the most important thing since the discovery of fire. My research topic had been in the field of cosmology—the study of the universe as a whole—and as a result my view of things tended to be somewhat excessively godlike. It was hard to pay attention to anything as hopelessly mundane as a brand new kind of star.

But there was a further reason for my general lack of enthusiasm, and it throws light on the mind of the scientist. I was not impressed that neutron stars had been discovered because I already knew they must exist. I had already read of the original arguments of Baade, Zwicky, and Landau and I had found them persuasive. It was not my fault that the observational astronomers—the ones with the telescopes, the ones who actually had to go and find the pesky things—had been so slow at getting on the job.

To a certain kind of mind it is almost as important to think of something as to discover it. To this sort of mind, a thing that has been thought of but not yet found nevertheless has an enormous reality. To such a mind, pure abstract thought is vitally important, and it is as real as tables and chairs. Mine is such a mind. To me thoughts and ideas have great significance. Also, I find thinking enjoyable and I do it all the time. I do it for fun. Some people do crossword puzzles, others dive down to the model railroad in the basement. Me—I think.

Such a point of view is common among scientists. It is, however, not universal. Other scientists had found themselves positively dumbfounded by the announcement of the discovery of pulsars. As recounted in Chapter III, Huguenin was such a one. Huguenin is relatively unpersuaded by abstract arguments. To such a mind as his, theories are all very well, but a single hard fact is worth a thousand words. Huguenin had certainly been aware of the idea of a neutron star prior to 1968, but he had paid it little attention. He had neither accepted it nor rejected it—he had simply filed it away. Perhaps it is significant that Huguenin had turned to the observational side of astronomy while I, by temperament, had turned to the theoretical.

At any rate, during the following year several more pulsars were discovered, and the debate raged as to what they were. But I was concerned with other things. I completed my dissertation, took a long and enjoyable vacation, and began a post-doctoral fellowship.

The institution of the post-doctoral fellowship is almost universal among scientists nowadays, and it is a good one. Most scientists do not earn their living by doing research. They do it by teaching. But teaching is not easy, and especially at the beginning, when one is learning the trade, it requires a good deal of time. The problem is that the same is true of research. The Ph.D. research project is usually the first a scientist ever undertakes, and it is not enough to ground him in all the techniques. Ideally, he would need some more time free of teaching duties to establish himself firmly in the field. This time is provided by the post-doctoral fellowship.

My thesis advisor was A. G. W. Cameron, and he had been everything a student could ask for. I was therefore particularly pleased when he offered me a fellowship, and I was happy to move to New York City to take it up. For the first several months I worked with Cameron on a project related to my thesis research. We had thought of something which might modify my earlier conclusions. This project led nowhere, for after a good deal of muddling around we realized that our new twist would have no appreciable effect. I tried to think of other research projects in the same general area, but nothing exciting suggested itself. Meanwhile, Cameron was urging me to work in some totally unrelated field in order to broaden my background. I did a piece of work on gravitational radiation. I was beginning to get bored.

Then the Vela Pulsar sped up.

Even today, as I write these lines, a tickle of excitement runs through me as I recall the moment I learned of it. It was a Sunday in my apartment in New York. Windows gave out over a sea of black-tar rooftops. Beside the easy chair lay a pile of unread *New York Times* accumulated over the week. Idly I was thumbing through them. An article caught my eye: ASTRONOMERS FIND A "CRAZY" PULSAR. As I read it a sense of astonishment filled me. I could not believe my eyes. I have always by temperament been attracted to the outlandish, the incomprehensible, and for some reason—a reason that to this very day I am unable to explain—Vela's behavior struck me in a way that nothing else about pulsars ever has. My first thought was that the *Times* must have gotten it all wrong. But the *Times* did not make mistakes like that.

It was a sign.

Cameron was in when I got to the office the next day. "For God's sake, Al, what's all this about dropping planets on pulsars?" I asked.

The article had quoted him as saying that something might have fallen onto the pulsar, speeding it up by giving it a glancing blow. All perfectly true. But I had pulled a used envelope out of the waste-basket and on it calculated that to give a big enough blow the object had to be bigger than Mercury. It made no sense: when was the last time a planet fell into the Sun?

Cameron had done the same back-of-the-envelope calculation. He leaned back in his chair. "The first I had heard of all this was when the *Times* called me up with the news. They asked me what might have happened and that is what sprang to mind." He permitted himself a small smile. "But upon more mature consideration I think it might have been due to the sudden onset of turbulence within the star."

Cameron is like that. He is a tall man, somewhat broad, but more than that he *acts* large. Watching him in a group is like watching a ship pick its way among a throng of rowboats. He is vast, formal, imperturbable. Cameron's is a deliberate, sober style of speaking, even in conversation: it masks one of the quickest, most inventive minds I have ever known.

He sketched out his latest idea. It involved a sudden transition within the pulsar from a smooth rotational motion to a chaotic, churning flow. I went off to my desk and thought about it. I came back and asked him some questions. Why Vela? Why pulsars and no other kind of stars? We went over things. In the afternoon he wrote up a short paper outlining his notion to be sent to *Nature*, and possibly because his thoughts had been joggled by some of my questions, he added my name to his as a coauthor. This simple gesture served even more to get me into the field of pulsars. And perhaps this, in fact, is why he so kindly allied my name with his explanation: he wanted me to begin working in some new area, and he thought this might lead me in that direction. It is the sort of generous thing that would occur to him.

At any rate, I resolved to learn everything I could about neutron stars. I started reading.

Months passed. Rumors began circulating about town. So-and-so had been on the phone to such-and-such, who had said that someone (he forgot who) out at the Jet Propulsion Lab was observing the pulsar. Its increase in slowing-down rate seemed to be gradually wearing away. Vela was slowing more slowly.

Cameron's idea had purported only to explain the speeding-up. Spurred on by the rumor, I began to ask myself whether anything in it could be used to understand these additional features of the event as well.

But as I turned the question over in my mind, a strange thing would happen. A peculiar tiredness would overcome me. I would grow unnaturally weary. At first I hardly recognized this strange reaction on my part, and I would briefly mull over the question and then pass on to other things. But as the question kept returning to my mind, so did the lassitude, and ultimately I recognized its presence. From this it was a short step to a recognition of its cause. I had been hiding something from myself. I had been avoiding the uncomfortable admission that I did not really understand Cameron's theory at all.

It was time to take the bull by the horns. Whenever I have a difficult piece of thinking to do, I am in the habit of doing it on foot. I take a stroll. This particular day was warm and pleasant and I set off for a nearby park. Once there the work began. It was hard. Somewhere in his explanation there was a point that I did not understand, but try as I might I could not put my finger on what it was that was bothering me about Cameron's idea. I could not even explain to myself the nature of my confusion. There was simply some point at which my mind stopped, as if it had encountered some invisible hurdle. Clarity of thought gave way to confusion, and I became muddle-headed and vague. When thinking so intensely I conduct arguments with myself, one half of me taking one side of a position and the other half the contrary, but this time I found the one half of myself simply irritated at the other that so adamantly refused to get down to work. Over and over again I would try to push the problem through. Over and over again I would return to some point of certainty, some matter of which I could be sure, and then take it forward, step by step, speaking to myself like a mother to her recalcitrant child. All the while I was walking, but only at the pace of a snail. I would take three steps and then stop, staring sightlessly at a tree, then turn abruptly and head off in some completely different direction. My mind kept wandering.

By noon I had gotten nowhere and it was a relief to give up the fight. I returned to my desk. Cameron was in and we ate in a local sandwich shop. During lunch I put the question to him. Cameron is fast. He immediately began talking. Like many scientists, he thinks out loud. As I recall, he seemed fairly satisfied with his explanation as it developed. But to me it was a mess. He had electrons moving this way, and magnetic fields moving that way . . . I could not figure it out. Soon I stopped listening. The sense of obscurity, of hopeless confusion, returned. I felt encumbered in fluff. It was difficult to breathe and I felt a desire to gasp. And then, inside my skull, something happened. I actually felt something *move* within my head. It was a physical sensation of something turning, like a boulder that I had been trying to roll that has suddenly broken free. I knew the explanation.

I burst in on Cameron as he was continuing with his ideas, and rapidly sketched what I had seen. He waited me out and then resumed speaking, and it soon became clear that he was acting as if I had never spoken. I butted in again and asked why he discounted what I had said.

"Well," he replied, "I didn't really understand it."

I ran through it all again, more carefully this time. After I had finished, Cameron said nothing for a while. He munched on a pickle. He stared off into space. There was a silence. And then he paid me the highest compliment one scientist can ever pay another.

"Good," he said. "That's it."

We returned to the office, and I was fairly bubbling over with elation. I felt as if a weight had been lifted from my chest. I immediately sat down to work, and began putting the insight into detailed form.

Sad to say, in the long run my wonderful idea has fallen by the wayside, for it was based on Cameron's set of ideas and time has shown these ideas to be wrong. By now both his and my explanations have become without significance. But no matter. It was fun while it lasted, and the moment at which I won through will always remain with me as one of those instants in which the pleasure of being a scientist is most keenly felt. Such moments are few and far between, and no one can order their coming, but when they do arrive they are precious, and they make it all worthwhile.

There were other rumors circulating about town. Malvin Ruderman also had a theory to account for the event that had occurred within the Vela Pulsar. Not only that, he had given it a name. He called it a glitch.

What is a glitch? Electronics buffs will recognize the word. It is slang. If you have built a new and delicate piece of electronic equipment; if this equipment has been working perfectly well for months now; and if suddenly, inexplicably, and completely without rhyme or reason it goes haywire—then that is a glitch.

Ruderman's theory of the Vela Pulsar glitch involved earthquakes occurring on the star, and the notion struck me as madness—wonderful madness. It appealed to my persistent love of the unusual. Because he was also in New York, I went over to see him. Ruderman's writing style is intensely formal, and from reading his papers I had already grown a little intimidated. How could anybody know so much? But on meeting him my nervousness melted away. An infectious excitement filled him, a cheerful vigor. There was nothing overbearing about him. His was a brilliance that delighted rather than overwhelmed. Within minutes we were both talking at once.

He showered me with reprints of articles on neutron stars. I went home and read the articles. I discussed them with Cameron and went back and discussed them with Ruderman. I got a few rudimentary ideas of my own and I asked him about them. We talked and talked. Hours passed in a moment. I was falling in love. The more I learned the more captivated did I become by the power and the beauty of Ruderman's ideas, and ultimately it was this more than anything that cemented my interest in neutron stars. It seemed to me then that any field of study that could attract about it ideas of such wonder was one in which I could work with pleasure for years. And I was right.

Ruderman's theory of the Vela Pulsar glitch relied upon the fact that neutron stars possess a solid crust. The Earth too has a solid crust, and from time to time the Earth has earthquakes. Ruderman proposed that the Vela Pulsar glitch was the result of just such a fracture—a *neutron starquake*.

How could such a starquake make the pulsar speed up? It did so by suddenly allowing the neutron star to shrink. If I were to sit on a barstool and give myself a spin, I would turn regularly only so long as I held quite still. If I were to pull my arms in, however, I would increase my rate of spin. If I did it quickly I would glitch. Ruderman calculated that the Vela Pulsar glitch would have been produced had it suddenly shrunk by about an inch.

Why had the starquake occurred, and why had it been so very large? On the Earth quakes were occurring all the time, continually changing its shape, but they were never so large as to change the length of the day appreciably. Ruderman gave the same answer to both questions: because the pulsar was slowing down.

What is the shape of a neutron star? Stars are spherical. But not precisely so. A star or a planet will be perfectly spherical only so long as it is not rotating. If it is rotating, on the other hand, it bulges out at the equator. The diameter of the Earth, for instance, is about 25 miles larger at its equator than at its poles. And because it too was spinning, the Vela Pulsar had a slight equatorial bulge.

Up to this stage in Ruderman's theory there was no particular difference between a pulsar and any other body such as the Earth. It was here that the crucial difference entered: pulsars were slowing down. Every year Vela's equatorial bulge diminished.

If the Vela Pulsar did not have a solid crust, this steady change in shape would occur with perfect smoothness. But solids do not change their shape so simply. They resist deformation. The crust of the pulsar was attempting to maintain its original bulged shape, but the more the star slowed, the more inappropriate did this shape become. The longer we waited, the greater were the strains upon the crust. Eventually the strains became too great. The crust cracked. It gave way. It fell inward at its equator—by one inch. And the pulsar glitched.

So Ruderman explained it. In a multiple collaboration with three other colleagues he also worked out an explanation for the increase in the slowing-down rate after the glitch and its subsequent gradual decay. But by far the most important aspect of his theory lay in a different direction. The important point—and it had struck Ruderman the moment he invented his idea—was that if his theory was correct, then in human terms at least the Vela Pulsar glitch was unique: it would never come again within our lifetimes.

He arrived at this conclusion when he calculated the *expected interval of time between successive starquakes*. Within his picture they regularly recurred. The quake came and the pulsar glitched— but then the pulsar resumed its steady process of slowing down; and as it did so the strains began to build up again on the crust. Ultimately it would crack again. It would crack once the pulsar was rotating sufficiently slowly. But because Ruderman knew how rapidly the Vela Pulsar was slowing down he was able to predict how long we had to wait until the next event.

The answer he obtained was measured in hundreds of thousands of years.

The conclusion was that none of us would live long enough to witness the next Vela Pulsar glitch. Neither would our children. By the time it came along, several hundred thousand years would have elapsed, and our age and its concerns would have entirely passed from memory. Very likely the human race would have progressed beyond an interest in things as childish as the pulsars. And who could even say that we would be here so very far in the future? If not, those faint radio signals carrying the news of the next event would fall upon deaf ears.

According to the starquake theory, we lived in a time of a gigantic and unlikely coincidence. We lived at the first epoch in recorded history in which the Vela Pulsar had glitched. If our race should not, after all, survive the next 100,000 years, then we lived at the *only* epoch in which any of its regular chain of glitches would ever be observed. It had occurred less than a year after the discovery of the Vela Pulsar. What are the chances that an event which could come at any time within a 100,000-year span actually does arrive so soon? They are one in 100,000.

It was a peculiar situation. Ruderman's and Cameron's had not been the only theories proposed. But Ruderman's rapidly came to the fore. The rest fell by the wayside. His seemed the most beautiful, the most logical theory; it held together and made sense. If it were not for that one flaw it would have been accepted without question. But the betting averages were against it—heavily.

It became a matter of opinion. Ruderman spent time working out various consequences and extensions of the theory—but he also developed a second, completely different explanation for the glitch. As for the rest of us, we lined up in camps. There were the quakers and the antiquakers. Believers found it possible to live with the 100,000-to-1 coincidence; the rest cast around for a better idea. As always in matters of belief, tempers rose. Once, while giving a lecture on the starquake theory, a proponent was crisply asked if he would trade the rest of his life for the time to the next Vela Pulsar glitch. His reply is not recorded.

So it went for several years. But then, in the fall of 1971, the Vela Pulsar glitched again.

6

How To Think of Something

We now jump forward five years.

In this period I had moved several times, found a teaching job, learned how to teach, and completed a number of research projects. Almost all of these projects were on pulsars; most were on the general set of ideas formulated by Ruderman. To these ideas I had myself (as well as many others, of course) been able to add a good deal. Also, in this five-year span the Vela Pulsar had glitched yet a third time, sealing the tomb of the starquake theory of glitches.

But most important of all, by the end of this period I had become tired of pulsars. I had worked on them too long. It is not a wise thing for a scientist to devote all his energies to one and only one field of study. Better to change fields from time to time in order to avoid getting stuck in a rut. The time had come to move on.

But before I did so there was one last bit of work remaining to be done.

This was a project studying the long-term evolution of pulsars. I wanted to take the best and most up-to-date picture available of the interior of a neutron star and see how it predicted a pulsar would evolve as millions of years passed. There was a certain amount of indirect observational data on this—studies comparing million-year-old pulsars with young ones—and I wanted to see if it could be understood by matching it with the predictions of the theory. But to be honest I must admit that this aim—the comparison of theory and experiment, which is supposed to be so dear to the hearts of scientists—was not why I took up the project. The real reasons were more personal. I decided to do it because I knew I could do it, and do it well; because I knew there was only a handful of people in the field as qualified as I. A further motivation was aesthetic, for this particular project would round out in a natural way my entire work on pulsars. It would dot the i's and cross the t's. Finally and most important of all, I was so sure I already understood the problem that I felt the research would go quickly and easily, and I would soon be free to change fields.

I was wrong on every count. The research did not go quickly and easily. It went slowly and painfully, and it developed into the longest and most difficult project I have ever undertaken. Nor did I understand the problem I had set myself to solve. As things turned out, I had completely missed its most essential feature, and I spent two years lost in a confusion because I had made so giant an error. And the project was not about the evolution of pulsars over geological ages at all. It was about glitches.

The physical system whose behavior I wanted to analyze consists of two components: the solid crust of the neutron star, and the superfluid lying deep within its interior. An analogy would be a glass filled with water. The pulsar is spinning in empty space. In the analogy we can accomplish this by placing the glass on a spinning turntable. Finally, although a superfluid has no viscosity, it does possess a weak "rubbing" interaction against the crust—similar to that of water against the glass.

The next element in the problem is that the pulsar is slowing down. It is as if we were to use a variable-speed turntable and smoothly reduce this speed to zero. One of the consequences of this deceleration is the steadily diminishing equatorial bulge of the star, but this was irrelevant to me. I was interested in something else: in its effect on the enclosed superfluid.

What is this effect? Step up to the glass on the turntable and grab it tightly. It stops in an instant. But the water does not. It keeps swirling for a while. In the same way, even if it is slow and steady rather than sudden, the act of deceleration forces the pulsar into a state in which its surface rotates more slowly than its interior. Miles down, ponderous, vast, millions of tons per cubic inch, the magical superfluid glides past the crust. It glides steadily, endlessly, in patterns that persist for centuries. We are describing a cosmic equivalent of the Gulf Stream.

How rapidly do these currents flow? If the rubbing between superfluid and crust is strong, they cannot flow very rapidly. If it is weak, on the other hand, they can and indeed do. This rubbing, in turn, is regulated by the temperature of the star. If the pulsar is hot, the rubbing is strong. The internal currents will then be weak. If the pulsar is cold, on the other hand, the rubbing is weak and the currents strong.

This is the problem I had set myself to analyze. A pulsar is born in the fires of a supernova explosion. Like the phoenix, it is born blazing hot. As the ages pass—thousands of years, millions of years— it slowly cools. And as it does so, the character of the currents deep within its interior slowly changes. They grow stronger. It was this gradual readjustment, requiring geological ages for its accomplishment, that I had decided to study.

One last point. It is a matter of common experience that rubbing liberates heat. We rub our hands together to warm them in the winter. Step on the brakes of a car and the brake shoes get hot. In the same way, as the superfluid currents rub against the crust of a neutron star, they heat it up.

The first step in solving the problem was to formulate it more precisely. Anyone could appreciate this general set of physical principles but I wanted to see what they implied in detail. Contained in this verbal description was the key to the understanding of an important physical phenomenon—but to achieve this understanding I would have to use some specialized techniques.

These were the techniques of mathematical physics. You begin with a set of general physical principles—those given above, for example—and write each as an equation. Then you solve the equations. I wrote down an equation expressing in mathematical terms the statement that the magnitude of the internal currents depended on that of the rubbing force between them and the crust. Another said that this force depended on the temperature. Still a third described the way in which the temperature was affected by the rubbing itself. My final set of equations was complicated: each was long, and there were a number of them. It took a good deal of space simply to write them all down.

In one sense I had achieved nothing at all at this point. There was nothing in these equations that was not also contained in the purely verbal description given above. The difference was that now I had before me a precise expression of the problem. I had translated a *physical* question into a mathematical one; from now on I could forget the physics.

Each of my equations was not a solution to a problem. It was itself a problem. It asked a question, and the answer to this question was the solution I required. As the last step I now had to find this solution.

But I could not do it.

* * *

To see what I was up against it is helpful to run through a few examples of equations and their solutions. The simple equation

$$2x = 8$$

asks whether there is a number which when doubled yields 8. Easy enough. We say that the solution is $x = 4$. But other equations are harder. Consider

$$x^2 + 2x = 8.$$

It is not so clear here what to do. One way, of course, is to guess: simply to pick a number out of thin air, double it, add the square, and check to see if we get 8. If so, well and good—but we are not very likely to guess correctly. Far better would be to find some means of generating the answer. This is how we solved the first equation, although the procedure was so simple it was largely unconscious: we divided 8 by 2. In a similar way mathematicians centuries ago developed a rule for solving the second equation: the quadratic formula. Application of this formula shows that there are actually two separate solutions, 2 and -4.

But the quadratic formula only works for quadratic equations. It is no help in the third example,

$$x = \text{cosine } x,$$

which asks if there is a number equal to its cosine. This equation is much more formidable than either of the above two. Here again we might conceivably be lucky enough to guess a solution, although it is not very likely. As before, what we really need is some method of generating it. But no one has ever succeeded in finding one. Generations of mathematicians have tried and failed. The third equation is asking a question to which no one knows the answer.

This is the problem which I now faced. My equations were far more complex than any of the examples, and for none of them did a method of generating the solution exist. In earlier ages I would have been forced to give up in despair at such an impasse. But not now.

I turned to a computer.

The equation

$$x = \text{cosine } x$$

cannot be solved by the methods of classical mathematics. But it can be solved on a computer. Indeed the solution is trivially easy, and with one day's instruction a high school student could write a program to accomplish what generations of mathematicians could not.

The computer solution proceeds by the straightforward expedient of brute force. For example, a particularly simple program could instruct the computer to choose some number—o, say—and check to see if it equals its cosine. If it does we are through. If not, the program instructs the computer to take another number—1, perhaps —and try again. And so it goes. The computer is only doing something that any of us could do: the difference is that because it is so fast it can succeed in carrying out this process, while a person would soon give up in disgust.

Here is a first stab at writing a program to solve the insoluble equation:

> Step 1: Set x equal to -1.
> Step 2: Add 1 to x.
> Step 3: Find the cosine of x.
> Step 4: Does x equal its cosine?
> If so, print out x and stop.
> If not, return to Step 2.

This sequence of instructions causes the computer to begin with the number -1 and (Step 2) add 1 to it. It obtains o. It then (Step 3) computes its cosine. If the cosine of o is itself o, the instructions (Step 4) cause the computer to print the answer and stop. Otherwise the machine returns to Step 2. Step 2 now tells the computer to add 1 to x, making $x = 1$. And the process continues for as long as we care to wait.

The instructions as they are written above are not expressed in any of the several computer languages now available. They are in English —our language, not the machine's. If one were to enter them into a computer, it would balk. But here is a translation of these instructions into Fortran, the most commonly used of the scientific programming languages:

$$X = -1$$
$$2\ X = X + 1$$
$$Y = COS(X)$$
IF (X.EQ.Y) THEN
  PRINT X
  STOP
  END IF
GO TO 2

A glance at this program is enough to reveal some of the language's peculiarities. The first program statement is simple enough, and sets

x equal to −1. It is the Fortran equivalent of Step 1 in the English-language program. The second statement corresponds to Step 2, and at first glance it looks like an equation to be solved. If so, however, it is a very peculiar equation indeed, for it is asking whether there exists a number which is equal to itself plus one. There is no such number, of course, but this need not bother us, for we have misinterpreted the statement. It is not an equation, it is an instruction. It is telling the computer to reset x to equal its old value plus one.

The number 2 preceding this instruction is not part of it: it is an address, and its function will soon become clear. Passing on, the third instruction corresponds to Step 3: the computer calculates the cosine of x and calls the result y. The fourth instruction, in turn, is equivalent to Step 4. The "if" statement

$$IF(X.EQ.Y) \ THEN$$

causes the computer to check whether x equals its cosine y. If the test is positive, the attempt has been a success; the machine then executes all those instructions up to the "end if" statement:

$$PRINT \ X$$
$$STOP,$$

which displays the value of X which solves the equation and then stops. If, on the other hand, the test is negative, the computer skips those statements and passes on to the next:

$$GO \ TO \ 2.$$

This returns it to the instruction whose address is 2—the one causing the entire process to repeat, with x replaced by $x + 1$.

As it now stands the program has some problems with it, and there is no guarantee that it will work. For example, it only tests positive numbers, and if the solution happens to be negative the program will never find it. This is easy to repair. Far more serious, however, is the defect that the program only tests integers. The solution may be some fraction such as 1.5: if so, the program will skip over it and continue blindly ahead forever.

This difficulty can only be met by writing a more sophisticated program. One possibility is to have the computer test whether each choice for x is greater or less than its cosine. For example, 0 is less than its cosine, but 1 is not. So the solution lies somewhere between 0 and 1. An appropriate program might guess pairs of numbers, each lying between 0 and 1, and each chosen such that the first is less than its cosine and the second greater. The program could then steadily whittle down the gap between the two members of this pair, increasing the lesser while decreasing the greater, and all the while checking to make sure that the solution has not been overshot. In this way the computer could steadily zero in on the solution.

Such a program will never find the precise solution we seek. It will only approximate it. For example, the choice $x = .99$ degrees is less than its cosine, but $x = .9999$ degrees is not. If we do not need an exact answer we can stop here, and say that the solution is close to .99. But we must always be aware that we have not completely solved the problem. This is endemic to computers and there is no way to avoid it: they can deal with equations entirely beyond the methods of classical mathematics, but never exactly. In truth, computers never solve equations. They "solve" them.

Faced with an insoluble set of equations, I set about writing a computer program to find an approximate solution. This turned out to be a very difficult task. The equations were complicated and so was the method required to solve them. I wrote and wrote. The logical sequence of operations employed by my program was tricky and it was easy to make mistakes. The computer was instructed to cycle through a given set of instructions some definite number of times. Sometimes there would be a subcycle which, depending on the results of the larger cycle, either would or would not do certain things. At one point there was a loop within a loop within a loop within a loop. There were instructions that went "if such-and-such is the case, and so-and-so isn't but if it was the case the last time you tried, and if you have performed this test an even number of times before, then do such-and-such."

After several months of writing it was time to type up the cards. The computer "read," not the written word, but punched Fortran cards: one program instruction, one card. I punched them up. The program consisted of about 2,000 instructions and the cards just filled a box. It was a hefty load to carry around. I ran them through the computer. The program worked.

Programmers know that if a new and complicated program works the first time it is tried, they are in serious trouble. It means their mistake is so subtle they do not even see that it is there. There are so many possibilities for error even in a short program that it is rare for one to work without revision. My program, enormously longer, had that many more opportunities to hang up. I stared glumly at the deck of cards.

I went on to test the program in every way I could devise. I strained it to expose its weaknesses. I ran it for high-mass stars and low-mass stars, for stars born exceedingly hot and those born relatively cold. I ran it assuming the superfluid currents beneath the crust to be absent—not because I wanted to know the answer, but because I had developed an intuitive feeling for the answer in this particular case. Finally I got a run in which the computer showed the pulsar's temperature to be less than absolute zero. I had found an error.

I chased the error down and fixed it. Now I had improved the program to the point that it would not run at all. I would feed the cards through the computer and it would immediately print out an error message—"program has attempted to divide by zero" perhaps, or "attempt to take the square root of a negative number." I was facing a bug. The term "bug" is slang for some small mistake, innocuous in appearance but catastrophic in its implications, that causes a program to produce errors like these. The worst thing about a bug is that you can be looking right at it without realizing its significance. A bug can be a comma in the wrong place, It can be the absence of a comma. A bug can be a typographical error such as a "23322" instead of a "22322," and although it is notoriously difficult for humans to spot such errors, the computer will respond to them every time. But in any case it almost always develops, once a bug is found and corrected, that the computer had been operating perfectly all along. It was the person that made the mistake.

My program consisted of 25 pages densely packed with Fortran commands. Somewhere within that morass lay my bug. I studied the program. I pored over it endlessly. I put in tests: instructions that caused the machine to print a message saying it had reached that particular point. Now when I ran the program it printed out the first few of these messages before stopping. In this way I pinned down the region of the program in which my bug lay. I went over the offending region with a fine-toothed comb. I got eyestrain.

Finally I found the bug. I corrected it. Now the program ran . . . but not for long. Now it hung up at some other point. I had cured the first bug only to run headlong into another.

Weeks wore by and turned into months. It was summertime now, and I promised myself that by fall I would be done. After all I had intended to dash the problem off. I was beginning to feel trapped. The months passed. I found the error. I found lots of errors. Summer ended and I was nowhere near finished.

Modern computers are operated from a terminal: a TV screen and a keyboard located in the comfort of your office—or your living room. It is a pleasant way to work. Not so back then, however. This computer read punched cards, and everybody using it was jammed together in a room. We fed our programs in through a card reader and the results came back on a printer. The reader gobbled cards at a frightening rate and it made a ghastly noise. The printer spewed forth page after page of computer paper in an instant. Also in the room were machines for punching up the cards. Programmers typed away, the machines clattered and thumped. There was a perpetual racket. People lounged around—undergraduates in torn blue jeans, graduate students with their kids. The floor was a mess. As for the computer itself, it was nowhere to be seen: off somewhere in another room.

Everyone knows the bitterness of the long haul. There is nothing more unrewarding than the slogging required to get a big program running. Meanwhile, someone had observed a strange new phenomenon having to do with galaxies. Interesting . . . but I was tied to my program. It was enormously cumbersome to use. I would make some trivial modification and it would have the most far-reaching effects. The program was a house of cards: pick one up and the whole structure collapses. I became angry. There were courses to be taught and committee meetings to attend. Students wanted to see me. Spring vacation came but I did not take it. By the next summer someone else had explained the new phenomenon about galaxies and covered himself with glory. Meanwhile, I was wandering through a maze of Fortran. Everybody cheered the launch of a new X-ray satellite. Everybody but me: I was busy, and hating every minute of it. A year had passed and I was facing another bug.

It was the strangest bug I had ever seen. By now I had improved the program to the point that it would appear to run correctly for a while. It was printing out the temperature, the rate of rotation, and the magnitude of the internal currents within the star. Furthermore, it was doing so not just once but over and over again, each time for a succeeding stage in the history of the pulsar. Last but not least, the solution obtained made good intuitive sense. According to the machine, the star was steadily slowing, steadily cooling, and steadily increasing the currents deep in its interior. It was all according to expectation.

But only at first. The program would run smoothly for a while but then go crazy, and it went crazy in ways that made no sense. Suddenly and without the slightest warning the steady cooling of the pulsar would reverse. The star would start to heat. At the same point it would start spinning wildly about. The computer was telling me that a major blast of heat was appearing within the star, as if some bomb had gone off deep within it. And accompanying this bomb the pulsar would suddenly increase its rate of spin.

Strange enough. It was summertime again, and with no teaching duties to disturb me I worked full time hunting down the bug. But I could not find it. I put in every internal test I could think of. The program passed them all. In desperation I caused the computer to print out every single subsidiary step it took. Seventy-five pages of printout, each densely packed with numbers, ended up on my desk. I took one look at them and went home early, lost in a black depression. The next morning I returned to the job with a heavy heart and commenced the task of making some sense of the bewildering morass. I went through it endlessly, looking for the bug. But I never found it.

There was no bug.

I do not know when it was that I first realized this. I do know I did not realize it all at once. There was no blinding flash of inspiration. Over all that summer, bit by bit, I came to realize that what the computer was saying to me made sense. As to what kind of sense it was, at first I could not even say. At first I only understood it in terms of the program—in terms of the actual sequence of operations I had instructed the computer to perform. Gradually I extended my understanding. Gradually I started to think, not in terms of the program, but in terms of the mathematics, the equations it was supposed to solve. Finally I began to think not of the mathematics but of the physics—of the physical system the equations described. It took me months to realize that the computer was telling me the truth.

I had discovered a new kind of bomb.

It was a bomb that only could operate deep inside a pulsar. It would go off by itself, and utterly without warning. When it did go off, it liberated an enormous burst of heat. And it made the crust of the star spin faster. *It made the pulsar glitch.*

The principle of construction of this bomb had been in my hands two years earlier, when I began the research. But I had not seen it. It is contained in the description of this project that I have already given. Had you, the reader, been on your toes, you would have seen it on reading that description.

The principle is twofold: (1) The hotter a pulsar, the stronger the rubbing between its crust and the superfluid currents deep within; (2) the stronger the rubbing, the more heat is generated by friction. Each of these statements refers to the other. They feed into each other. And so we have a feedback loop. In technical terms we have an *instability*.

Suppose that some outside agency were to heat the pulsar very slightly. Perhaps a fragment of a meteor were to crash into it, for example, liberating a tiny burst of heat. Or perhaps some small fluctuation in the magnetosphere were to warm its surface in the pulsar equivalent of a microscopic lightning bolt. What is the consequence of this utterly negligible extra heat? By statement number (1) it makes the crust rub a little more strongly against the internal currents. But by statement number (2) this in turn heats the star a little more. Now statement (1) takes over a second time, and the rubbing becomes still stronger. Then statement (2) a second time, and the heating more intense. So it goes, snowballing, accelerating . . . a runaway. That tiny initial burst has triggered a catastrophic heating of the star.

Such is the bomb. In comparison with it the glitch is a mere side effect. The glitch is the signal that the bomb has gone off. The superfluid currents spin more rapidly than the crust. The hotter the star, the more strongly they rub against the crust. They are working to speed it up. The burst of heat enables them to do it: it is they that cause the glitch.

It all comes back to the initial burst of heating. But what caused the heating? Was it the meteor or the lightning bolt that ultimately caused the glitch? An example of a second kind of instability will answer this question. Consider the problem of balancing a pencil on its tip. Such a pencil is unstable: it falls over the moment it is released. Of course if by some wild chance we could succeed in balancing the pencil perfectly, it would never topple. But this never happens. The most microscopic of events is enough to disturb the perfection of the balance we have achieved. An infinitesimal breath of wind blows the pencil sideways. A microscopic shudder of the table jerks it over. In truth we do not need to ask what disturbed the pencil's balance: anything, no matter how trivial or microscopic, is sufficient to send it tumbling. The important thing is not the initial disturbance at all. It is the simple fact of the existence of the instability.

The glitch could have been caused by anything.

So it was that I was first attracted to the study of pulsars by their strange and enigmatic glitches; that I spent years working on them, learning wonderful things—some not so wonderful—and finally by an accident, blindly, like a sleepwalker I stumbled on an explanation. Is there any moral in this story? Looking back upon it all, is there any method I can find for making still more discoveries?

I think not. The story is too complicated, too full of twists and turns, accidents and stray circumstances, for any simple moral to be drawn. It is just what happened.

Scientists are the most skeptical people in the world. As things now stand, most workers in the field believe the true explanation for pulsar glitches lies in some different direction from the one I found. I myself am not so sure about it. At question is the precise nature of the rubbing interaction between the superfluid and the crust, and whether it is really so sensitive to the temperature of the star. If not, the feedback loop I have described is broken and the theory tumbles to the ground. Superfluids are messy things; right now we do not know enough to be sure. There are some experiments which could be performed that might answer the question, but it is difficult to be sure whether they would. There are theoretical calculations that might help, but they too are hard, and themselves fraught with uncertainty.

I do not know if the theory is right. I do not know if it is a good one, correct in its essential respects—or whether it is a pipe dream: irrelevant. I cannot say whether the years of toil I expended in this work were justified, and if the task was worth the effort. Nor do I believe I ever will know for sure.

There is a deep and profound difference between the natures of observational and theoretical research. Here is an observation proving that the temperature of the Sun is such-and-such. Here is another demonstrating that the Andromeda Nebula is so-and-so many light years away. These are solid facts and we can depend on them. Theoretical research, on the other hand, does not lead to such certainties. Usually its results are quite intangible. It leads to new ideas—but ideas are uncertain and debatable. It leads to new points of view—but this is not enough if what we want is hard and fast results. It is only if we are very lucky that far down the road theoretical research leads to what we have been looking for all along: understanding.

I must confess that there are moments when the thought I might be correct fills me with a kind of terror. There are times when I cannot believe I could have succeeded in penetrating to the heart of this mystery. How can someone who has trouble keeping his bank account straight claim to know the secrets of the pulsars?

This research required the use of a computer. It could not possibly have succeeded without one. And this is true of many other research projects as well. In the space of a few short decades these machines have become indispensable to the progress of science. Everyone knows the space program would not have been possible without the development of new, ultrapowerful rocket engines. But it would also have failed without computers. No human or team of humans could control the complex sequence of events required to launch a rocket. No human could accompany a space probe on its way to Saturn and control its operations on arrival there. Computers do.

They are equally ubiquitous, of course, in daily life. Computers make airline reservations. They keep track of our bank accounts and send out phone bills. A doctor takes a blood sample and the lab tests come back in computerized form.

Many people are disturbed by this proliferation. I myself am not—after all, I use the machines all the time. But there is a danger in the growth of computers that concerns me. It arises from a common misconception they have engendered in the public eye. I believe that computers are dangerous because they convey an illusion of certainty about their operations. They are the most perfect machines that have ever been constructed. They almost never make mistakes. A computer can multiply 31.835521 times 14739.447 in one ten-millionth of a second and get the right answer every time. It can add one billion numbers together in a minute. But so what?

Computer programmers have a saying, "Garbage in, garbage out." By this they mean that the accuracy of a computer in and of itself is irrelevant. There was certainly nothing so flawless about *my* interaction with them. My story is full of all the false starts, blunders, and confusion that fallible humans are prey to. In the end the wonderful perfection of the machine led only to the invention of a theory—a hope, an idea.

No—it was not the computer that solved my problem. It was another machine: that small chunk of gray matter lying within my skull. Scientists have been getting along perfectly well without computers for centuries until recently, and if every computer on Earth were magically to vanish, I have no doubt that science would continue to flourish. The computer is only a tool. At its heart every account of how some discovery is made is the retelling of the same tale: the story of the workings of a creative mind.

Surely the workings of the mind are utterly mysterious, and they are hidden to us all. They are especially hidden to the mind that is doing the thinking. Thinking is not spectacular. Indeed, I have found it difficult to decide just when I was thinking and when I was not. It has been a source of endless astonishment to me to observe how often I have arrived at the solution to a problem at moments when I appeared to be concerned with something else. Many years ago I was faced with an unusually troublesome bug. I had spent more than a week on this one. By the end I had nearly succeeded in memorizing the program, but I felt no closer to finding the bug than ever. Ultimately I did succeed—while asleep. I was awakened by a nightmare. I lay in bed, mouth dry and heart pounding, and as I lay there I realized that I knew the error in the program. I did not find the error in that instant: I realized that I had found it already.

On another occasion, I was standing in the shower when an explanation for something occurred to me. I had not been thinking about pulsars. I had not been thinking about anything at all—I was just standing there lazily, appreciating the hot water, when it quietly crossed my mind that such-and-such an observation could be explained in such-and-such a way. On still a third occasion an idea occurred to me while mowing the lawn. Sometimes ideas come only after hours of painful thought; other times unbidden and by themselves. And in every case I have had the same curious feeling that it was not I who thought of the idea; but rather that the idea, wholly formed by some outside agency, finally floated into my awareness. It is as if the real activity of the mind lies entirely outside of our consciousness; that ideas are formed in darkness, and only become visible when they please.

Even though creative thought is crucial to the activities of the scientist, it is never taught in school. No one is ever taught how to think. In college and graduate school I took courses in a wide variety of subjects—physics, math—but I never took one in creativity. This cannot be taught. Luckily, it does not need to be. Creative thought is universal in the human species and everyone, scientist and non-scientist alike, thinks creatively all the time. Ceaselessly, incessantly, largely outside of our awareness and beyond our control, our minds ripple along playfully; comparing ideas, generating ideas, rejecting ideas. From time to time something bubbles to the surface. Creativity flowers.

Albert Einstein once wrote that the most incomprehensible thing about the universe is that it is comprehensible. For a long time I did not know what he meant, but now I believe I do. For after all, what was the nature of my research? I did not study any pulsars. I visited a radio telescope once, looking for the pulsar, but all I saw was the telescope. Neither have I ever seen a neutron or an electron.

But none of this prevented me from successfully carrying out a program of research on these objects. How did I do this? By thinking. And thinking, in turn, is introspection: it is the act of looking inside oneself. Not once did I look at anything beyond a book, a computer printout, or a piece of paper covered with markings that I myself had made. *Observational astronomy may be the study of telescopes, but theoretical work is the study of the contents of one's own mind.*

My thinking takes place in English—but no pulsar ever heard of English. I solved equations—neutron stars do not. I went to school, learning the physics—the natural world never needed to. The natural world simply is.

The elementary rules of logic, language, and mathematics: these are the ingredients of research, and they are strictly human things. They are inventions of our species. How can it be that their application tells us valid truths about the far-off universe? Here is an object: a brain. Electrical currents flow within it, chemical reactions take place throughout its bulk. It is made of protoplasm, proteins, DNA. Now here is another object: 1,000 light years away, ten miles in diameter, blazing hot, spinning wildly, superdense. In some way the pattern of activity of the first can be made to mirror, to mimic that of the second. This is the magic of creative thinking: the most powerful tool the human race has ever found.

PART TWO

———•———

BLACK HOLES

7

Fire and Ice: The Time Machine

It is a warm and cheerful day in early summer, and by midmorning the beaches along the California shore are beginning to fill. A woman lies sunbathing on a towel. Lulled by the warmth, she has almost fallen asleep. Beside her a portable radio murmurs into her ear. Children are shouting as they run along the beach. The pounding of the surf barely penetrates into her consciousness.

But now, roused by a familiar tingling on her shoulders, she sits up and automatically fumbles for the suntan lotion. Glancing about as she rubs it on, she finds that she is squinting, so brilliant has the sunlight become. Nearby a casual volleyball match has halted in mid-game by mutual consent, and the players have retreated to the shade of beach umbrellas or the coolness of the surf. Elsewhere a couple have gathered up their belongings and are walking to the car. Mindful of sunburn, she arranges a towel across her legs.

As she gazes about, the woman realizes that the glint of sunlight as it dances on the waves somehow seems to have changed. It has become more piercing than usual: bluer, harsher. It hurts her eyes. She holds out her arm, and its shadow on the sand stands out in hard, sharp contrast. Unaccountably, the colors of her beach bag have become unfamiliar to her, and those of the houses lining the shore seem strange. Even the greens of the hedges have changed. The familiar coloration of sunlight, usually a warm yellow, has gradually changed to a harsh electric blue.

She rises and, with a sprint, dives into the water. Overhead the Sun glares down. It is a little smaller than it ought to be.

Now the first half hour has passed, and the beach has been entirely deserted. The few remaining bathers are nervously preparing for the dash to the safety of their cars. The sunlight has become vicious, a blinding and ferocious glare of blue. Several hundred miles to the east, in Arizona, a vacationing family is driving across the Mojave Desert en route to the Grand Canyon. On the flat expanse of desert there is not the slightest shade, and they feel as if they were trapped on the surface of a broiling pan. Without saying so, the father at the wheel has become worried, for although he is wearing dark glasses he can barely see. Squinting nervously, covering with his hand first one eye and then the other, he is fighting to keep the car on the road. His wife beside him has closed her eyes, and the children in the back seat are silent and subdued. In the shimmering blaze of light he barely notices at first the painful burning sensation on his arm. He is resting his elbow out the window and it is exposed to the Sun. With a sudden grunt he pulls it inside. They pass a sign and he dimly becomes aware of what it had said: "Kingman—20 miles."

Kingman . . . there will be shade there, he knows, and he steps on the gas. Hoping the tires will not blow in the heat that is intense even for the desert in midsummer, hoping that the radiator will not boil over, he rounds a bend in the road with tires squealing. Is that a truck ahead bearing down on them? He momentarily closes both eyes, then snaps them open for a better look. Yes! Although he is not aware of it, he has a headache now, and the children are crying. A tire goes off the road into the dirt. With a jerk he pulls the car back into the lane. The truck passes with a roar. Briefly he closes his eyes. The next bend in the road he does not see.

One hour after it had begun collapsing, the Sun has shrunk to half its normal size. Its progress is steady and unhurried. Now its glare is blinding. Soon it is a little brighter, and a few minutes later brighter still. No thunderous booms rain down upon the Earth from the dying Sun. The slow destruction it wreaks is accomplished in utter silence. Flowers still nod in the garden, babbling brooks still run in their courses, and the gentle breezes still blow. It is only the terrible light of the Sun that has changed. A woman emerges from the subway in New York, glances up in shock, and makes a quick turn into the nearest office building. A man behind her halts irresolutely, and then returns to the subway. Drivers pull their cars over to the side of the road, anxiously glance about, and make sudden dashes to shelter.

On a pleasant street in the suburbs, a boy stands in the overwhelming sunlight, eyes squeezed tightly shut, completely immobilized. His skin is burning. Panicked, he opens his eyes for a moment, spots a nearby bush, and worms under it for the protection of its shade. Ten minutes later even this is not enough. He begins to cry uncontrollably. Not far away, indoors, a man stands listening in his living room. His throat is dry and his heart is pounding in terror. Try as he might, he simply cannot bring himself to go out of the safety of his house to rescue the boy. He has drawn the window shades, but even so around their edges the sunlight pours: dangerous, annihilating. The cries outside have halted now. With a sudden woosh the roof of his house bursts into flame.

Over the entire sunlit hemisphere of the Earth, people trapped outside die slowly of hideous burns. Those indoors die when the buildings ignite. Now the forests flash into flame. Birds tumble from the sky. Ninety minutes after it began its collapse, the Sun has shrunk to a fiery point of light in the sky. Lakes, rivers and oceans smolder and boil. Clouds of steam mingle with the acrid smoke of burning cities and obscure the final destruction. Down through the choking haze the Sun pours its ravaging glare. It gathers, and leaps in a final overwhelming flare. In these last instants the very surface of the Earth itself begins to liquefy, and flows in streams of molten rock.

In London, the Sun had set an hour before, but now the western sky is briefly illuminated by a new and frightening light. Close upon its heels comes a scalding storm of superheated air and steam, as the very atmosphere of the Earth pours from the sunlit to the darkened hemisphere. Stronger than the hurricane, boiling hot, covering a band thousands of miles wide and stretching in a ring entirely around the globe, it topples buildings and blasts airliners from the sky. A single sharp burst of gravitational radiation, a last gravitational scream from the Sun, shakes the very fabric of the Earth. The planets wobble briefly in their orbits.

And then blackness.

The Sun is snuffed out. In an instant, in one ten-thousandth of a second, its terrible brilliance is extinguished. Now, as the smoke and steam slowly clear from the sky, the stars come out and shine down upon a scene of unparalleled desolation. Where once had stretched broad fields, forests, and lakes, there now appears a scorched wasteland as barren as the surface of the Moon. Not a bush, not a tree, has been left standing. Gentle meadows have been baked into harsh plains, and forests into heaps of ash. The very soil has been fused into a hard, lavalike consistency. Entire cities have been obliterated, and every wooden home in North and South America is gone without a trace. The bulky skyscrapers of New York City have been reduced to heaps of slag. The Statue of Liberty is a pool of molten metal. Clouds of ash, blown by the wind, filter down upon the skeletons. The Great Lakes have entirely boiled away, and what was once the broad expanse of the Mississippi River is now a dry wash meandering past dead hills.

From start to finish, a mere two hours have passed.

At this very moment, in an apartment in Tokyo, a man lies sleeping, protected from the Sun by the vast bulk of the Earth. Momentarily disturbed by a mild shudder as the burst of gravitational radiation passed by, he had soon returned to sleep. Hours later he begins to toss and turn. Eventually he sits up in bed and looks about him. Outside it is still pitch black. Stars shine in through the open window. Lying down again, he closes his eyes, but unaccountably finds himself unable to fall asleep. Bit by bit into his consciousness a peculiar and nagging fact is penetrating: although it is the middle of the night, the street outside is clogged with traffic.

Finally, wide awake now, he wraps a bathrobe about him and pads into the kitchen. Turning on the light he glances at the clock. It is nine o'clock in the morning.

Unbelieving at first, he simply assumes that something has gone wrong with his clock, but soon a note of panic in the babble of voices from the street below persuades him that something indeed has happened. Switching on the radio, he begins to dress, but keeps pausing in astonishment at what he hears. Reports of terrible storms in Eastern Europe and the South Pacific Islands . . . all communication with North America inexplicably cut off . . . the JAL flight from London hours overdue . . . and over and over again the incomprehensible and incontrovertible fact that the Sun has failed to rise. But what sets his heart to pounding in his chest is not the darkness outside, nor is it the news that the announcer is giving. It is *how* the announcer is reading the news: it is the note of pure and rising terror in the announcer's voice.

The man hurriedly finishes dressing and descends to the street. And only there, jostled by the crowds, swept helplessly off his feet from time to time by surges of panic-stricken people, does the enormity of what has happened finally begin to penetrate. Somewhere, someone is screaming.

In Moscow it had been a cool night, and now no rising Sun comes to dispel the chill. A woman, the wife of a minor government official and the mother of three, closes the windows of her small apartment and rummages in the bureau for a light sweater. The children are at school (though God alone knows what the teachers will be saying to them, she reflects) and, being a practical person, she decides to stock the larder as a precaution. Holding down her rising flood of fear, she battles through the terrified crowds to the store. There the crowds are thicker than ever, and the shelves even more bare than usual, but she manages to find much of what she needs. The store-keeper is besieged by shouting customers, and when she finally reaches him he demands twice the usual price. Uncharacteristically, she accepts without a murmur, and begins the long walk home. By the time she arrives she is utterly shaken, and the oppressive darkness has begun to tell on her. She sits down and with shaking hands lights a cigarette. Outside, clouds are gathering. A cold rain begins to fall. She rises, opens the closet door, and begins hunting for the winter clothing.

Riots in Singapore . . . terrified crowds in Bombay . . . a brief power failure in Jerusalem. Overhead the imperturbable stars wheel slowly past. "Noon" comes. Still no Sun. "Evening" comes, and now the first "day" has passed. Throughout half the world the destruction is complete. Throughout the other half incomprehension and shock slowly give way to panic. At first it is the endless darkness and the incomprehensibility of the event that weigh upon the spirit. At first no one notices the gathering cold.

Within a matter of weeks the first snows begin to fall in Egypt. Palm fronds whip back and forth in the icy blasts. No mere few inches of snow accumulate, but yards and yards of it: those vast quantities of ocean water, boiled away by the heat of the collapsing Sun, are returning now to the Earth in one last, long snowfall. Alexandria is paralyzed by the first snowstorm in its history. The Sphinx gazes enigmatically across the drifts. The Nile freezes over. Now it is "midnight." Now it is "noon." The cold deepens.

In the hotter, equatorial nations of the globe, entire populations are wiped out almost immediately by the unfamiliar cold. In the harsher regions, where central heating is a way of life, people survive a little longer. Twenty degrees below zero in Geneva. Trucks making deliveries of heating oil are hijacked at gunpoint. Grocery stores are ransacked by marauding crowds. Fifty degrees below zero. Water mains freeze and then burst. Mighty Orion shines down, glittering in the iron cold. Traffic is immobilized by giant drifts of snow, and deliveries of food to the markets come to a halt. Suddenly the city is plunged into darkness as the electric power blacks out, brought on by unprecedented demands for light and heat. With superhuman effort the engineering staff manages to restore power. One week later it fails again—for good, this time.

In a Peking apartment equipped with a fireplace, twenty people gather. Bit by bit, they are feeding the wooden furniture into the blaze. Soon it is exhausted, and someone bravely ventures for more into the deserted apartment next door. People huddle around the fire, endlessly gazing into its warm heart with anxious eyes. Outside, the snow has drifted so deeply that the door has been blocked shut. They are trapped inside. Not far away a building has caught fire, and crowds are gathered around the blaze. Soon the building next to it catches fire too, and then the next. Warmth! No fire engines come to extinguish the conflagration. Floundering through the snow, more crowds gather to warm themselves near the flames.

One hundred degrees below zero. Now the cities of Europe, Asia, and Africa are silent, lifeless tombs. Only here and there, isolated, huddled in the dark wilderness, small bands of people have managed to survive. None of them is aware of the existence of any of the others. Each believes themselves to be the last survivors of the human race.

Two hundred degrees below zero. Eventually, everyone succumbs.

Far below the surface of the few remaining oceans, the cold at first does not penetrate, and the absence of sunlight is nothing new to fish accustomed to eternal darkness. But eventually the ocean surface itself freezes over. From outer space, dimly lit by the stars, the Earth is utterly featureless now, a uniform sphere shrouded entirely about in deep blankets of snow. The familiar outlines of the continents are obliterated, and the ocean basins filled with ice. Finally, unseen, far below the surface, the last remaining drop of water freezes and the last fish dies. Life has come to an end on the planet Earth.

Now the very atmosphere itself begins to liquefy. A new rain falls, a rain of liquid air. Briefly, it flows in bubbling icy brooks and gathers into clear ponds. Then these too freeze solid. The wind blows more and more thinly, and soon ceases altogether. It has been frozen out of existence. The terrible vacuum of interstellar space penetrates to the very surface of the Earth.

Everything stops.

Before the Sun vanished, the planets of the Solar System had swung about it in their orbits, forced to do so by its gravitational pull. They are still swinging in these orbits, but now they are moving about a black hole. The above account is a more or less accurate description of the course events would follow if the Sun, suddenly deprived of a means of supporting itself against its own weight, were to collapse in on itself. If a gas is compressed it heats up, as witnessed, for example, by the experience of pumping up a bicycle tire and finding that the pump itself has become quite hot. So it would be with the gases comprising the Sun: the intense heat generated by the collapse would have devastating effects upon the Earth. By everything we know such a fate is not, in fact, in store for us. The Sun will end its days more serenely. But this will occur to other stars. It already has.

But all this was only the beginning, only the preliminary phase of the collapse. In the final phase something quite different had occurred, something unique in all of physics and astronomy. The Sun had been transformed into a black hole.

In contrast to a neutron star, which surrounds itself with violent, stormlike activity, a black hole surrounds itself with silence. It is entirely still, and in a very real sense it is invisible. Move up to within a few hundred miles of the black hole that once was the Sun. Gone is the brilliant sunshine, the solar wind, the dazzling solar corona. On first glance now there is nothing to be seen at all. Nothing to be seen, perhaps, but much to be felt, for there is an intense tug of gravity. It is as if nothingness could exert a force.

The distant constellations are strangely altered. Ahead lies a small dime-sized patch of sky entirely devoid of stars, and around it a distorted image of the pattern of the sky. The black hole is drifting sideways. If a black *sphere*—a rock—were to drift by, one would see a silhouette moving across the backdrop of the stars. But this is different. This silhouette does not simply cover things up: it expels them from a circular region of the sky. As the empty disc approaches it, the image of a star slithers smoothly upward out of its way. Once the silhouette has passed, it gently slides back down again. A star lying below the path of motion moves downward out of the way. Finally, a star lying exactly on the path does something more spectacular— its image briefly splits into a brilliant circle of light ringing the hole, and then coalesces into a starlike point on the opposite side once it has passed. A black hole is a gravitational lens, and the distant sky ripples about it.

Back away now from the intense attraction of the hole, and return to more prosaic regions of space. While we were close, something strange has happened. The rest of the universe has aged a little more than we have. If we have spent ten minutes examining the gravitational lens, ten minutes and one second have passed farther out. The black hole has sent us into the future a little more rapidly than usual. A black hole is a time machine.

In normal life all of us are steadily and inexorably marching into the future, each at the same rate as all the others. The structure of everyday experience is such that this march appears to be absolute, unaffected by anything we might do, and universal. We all believe, and experience confirms, that if ten minutes have elapsed in one place, then ten minutes have also elapsed everywhere else. It may be noon in Philadelphia when it is 9 A.M. in Los Angeles, but we understand this to be a trick induced by the fact that the Earth is round. It is not a fundamental property of time. What is fundamental is *elapsed* time. If I am speaking to you by phone from Philadelphia and you are in Los Angeles, it is abundantly clear to us both that time is passing equally.

The black hole destroys all this. If I were to venture close to a hole and you were to remain farther out, I would observe you to be breathing at an unnaturally rapid rate. To me your gestures, your speech, the very beating of your heart would be hurried and excessively rapid. The clock beside you would appear to be ticking far too rapidly. You, on the other hand, would claim that I was moving in slow motion. But it is not the clocks, it is not our bodies that have been affected—it is the very passage of time itself.

Drop closer to the hole. Now it fills a sizable portion of the entire sky. Still there is nothing to be seen. The very term "black hole" is not appropriate, for it implies something black that can be observed. A wall painted jet black can still be seen but a black hole cannot. It looks like empty space looks. A flashlight pointed at it reveals nothing whatsoever. Nor does it show up on a radar screen. If it were not for the tug of gravity and the radical distortion of the distant constellations, there would not be the slightest evidence of the hole's existence. There is no sense of slowed time. Wait ten minutes and then back away. Eleven minutes have passed outside.

Venture still closer to the vicinity of the hole. Half a mile beneath is a vast floor of invisibility. It is as black down there as the inside of a cave. Above the stars are a harsh, brilliant blue and distorted images of one's body hover on every side. Every ten minutes, twenty minutes elapses up above. The force of gravitation is gigantic now, and the farther we drop the stronger it becomes. Move still closer. Drop to within one foot of the unseeable edge. For every ten minutes, seventeen hours flow by overhead. The mightiest rocket engines NASA has ever built would not be able to lift a pea down here. Once more drop down—drop twelve more inches. . . .

The Sun is not far below now, and it is falling inward upon itself. It was centuries ago that it had vanished, leaving the Earth to freeze, and all that time it had been there, collapsing at almost the speed of light, and in all that time it had fallen inward not even an inch. We are falling too now, helplessly, but it falls out from under us. In a fraction of a second it has collapsed to a point one mile beneath our feet. Toward this point we plunge headlong. Above us other things fall into the hole, and rain down upon us—planets, stars. The vast bulk of the Andromeda Nebula shoulders in. Everything is in motion now, everything falls. Shout for help and the shout falls downward and inward. Send a powerful burst of radio waves up, and it falls down. Shine a light straight out, and it falls straight in. Nothing leaves a black hole. But before we have time to think of these things, we have fallen to the center of the hole. Overwhelming forces take control, and we are crushed out of existence.

Newton, Einstein, and Schwarzschild

A black hole is a pathology of gravitation, a singularity in an object's gravitational field. As for the nature of this object, it could be anything—a star, a planet. The object is not important. It is the singularity that counts. This singularity wraps around the object that made it and renders it wholly invisible. It bends the paths of light rays. It distorts the image of the distant sky. It distorts the very fabric of time and space. And finally, it reacts back upon that object and crushes it out of existence. This destruction is hidden behind a veil. Before our very eyes a solemn event is being enacted—the drama of the ultimate fate of matter. We never see it.

It is easier to say what a black hole is not than what it is. It is not a hole. "Hole in space" is a phrase occasionally used, even by scientists who should know better, to describe them. But it is not a good description. More than that—it is meaningless. A hole is a place where there is no matter; space is a collection of places where there is no matter. The phrase has no more sense to it than "a space in space" would.

Actually the term has quite a different meaning. It is slang. In England a black hole is a *jail*, as in the black hole of Calcutta. In this sense it is good enough: the black hole is a prison in which matter is entombed forever.

In approaching so strange and forbidding a subject, it is best to start on familiar ground. Return to something we have already done —to the imaginary experiment conducted in Chapter 5. In that experiment a cube of rock was crushed to higher and higher densities, and we were interested in following the various changes that occurred in its internal constitution as it was compressed. But now crush something bigger. Crush the Sun itself.

In doing this we are only repeating the catastrophic collapse of the Sun described in the previous chapter. The difference is that now we will do it slowly. There will be time to look at things in more detail. As the Sun is compressed it undergoes the very same series of transformations that was described in Chapter 5: it passes from a solid to a superfluid to a mix of elementary particles. But although all this happens, it is not the point. The point is the force of gravity on the surface of the Sun.

The Sun is very massive and the force of gravity it exerts is large—much larger than the Earth's. If a 150-pound man were to stand on the surface of the Sun he would weigh a full two tons there. And if we were then to compress the Sun, his weight would increase still further. As the Sun is compressed the very same quantity of matter is being packed into a smaller and smaller region of space and the force of gravity it exerts grows stronger. The smaller the Sun becomes, the more that 150-pound man standing on its surface weighs.

Also important is *the velocity of escape* from the Sun. This is the velocity that must be imparted to an object in order to send it flying away into space. Escape velocity from the Earth is 7 miles per second. Throw a stone any slower and it falls back to Earth. Throw it at any greater velocity—by firing it off in a rocket, for example—and the stone is launched to the stars.

Escape velocity from the Sun is a good deal greater, of course. It works out to about 390 miles per second, and if the Sun is compressed it grows yet larger. If it is shrunk to just half its present size, the weight of the 150-pound man is increased to eight tons, and escape velocity to 550 miles per second. At one-tenth the present radius, the man weighs a full 200 tons, and the velocity of escape has climbed to more than a thousand miles per second. If the Sun is compressed until it is just the size of the Earth, the enormous concentration of matter attracts the man with a force of 25,000 tons, and escape velocity is 4,000 miles per second. Still more compression. Crush the Sun down upon itself until it is a mere 1.75 miles in radius. Now gravity has assumed such an unthinkable intensity that the velocity of escape is 186,000 miles per second.

186,000 miles per second . . . that is the velocity of light.

In the year 1905, when he was twenty-six years old, Albert Einstein received his doctoral degree from the University of Zurich. In the very same year he published three papers in the German journal

Annalen der Physik. Each one of these papers stands as a major landmark of physics. The first developed a theory of Brownian motion: a theory that provided the final decisive confirmation of the atomic nature of matter. The second developed a theory of the photoelectric effect, and in so doing took a major step forward to the creation of quantum mechanics. And in the third paper Einstein presented the special theory of relativity.

Over the next decade Einstein extended the ideas of special relativity into new domains. The result was the publication, in 1916, of the general theory of relativity. The full theory of relativity was then complete, and to this day it stands as one of the grandest of all the creations of the human race.

It is the mark of a bad theory that it continually requires modification and readjustment in order to accommodate itself to new discoveries as they are made. Theories like relativity are precisely the opposite: the new discoveries conform themselves to it, and they do so in ways that the inventor of the theory could not possibly have foreseen. The theory turns out to be more true than its creator had thought. In 1905 Einstein had predicted that time passes more slowly for an object in motion than for one at rest. When he made this prediction there was not the slightest practical possibility of testing it. It was not until several decades had passed that this possibility arose and Einstein's prediction was confirmed. As it was ultimately performed, the test involved studying the decay of an unstable elementary particle in motion, but in 1905 Einstein had never heard of this particle. Neither he nor any other physicist of his day had the slightest inkling of its existence. In some way his theory had been so profoundly in accord with reality that it jibed with discoveries lying far in the future.

Special relativity had also predicted that the mass of an object in motion should be greater than when at rest. Today this principle is a truism to the engineers who design giant elementary particle accelerators, but in the year 1905 such accelerators lay generations off. They were inconceivable. Galaxies too were inconceivable in those days, but when finally they were found, the vast and majestic expansion of the universe was revealed. It had been predicted by Einstein.

Einstein achieved all this without massive federal funding, without hordes of assistants, without telescopes and computers. He did it by himself, using only the powers of his mind. He did not rely on the latest data. He did not read all the journals. Instead he retreated into his own solitary thoughts, and he decided how the physical universe ought to behave. Out of pure *a priori* reasoning he constructed his theories and he made them work.

In surveying the progress of science, nothing prepares us for Einstein's achievement. His work was not a logical outgrowth of the methods of his day. Einstein invented new ways of doing physics and he thought in ways utterly unlike those of his contemporaries. The more his age recedes into the past, the greater he becomes. Even as a young man Einstein was regarded as one of the best physicists of his day, and as he grew older he grew in stature. By the time of his death he was held in awe. And from our present vantage point he towers over his age as Shakespeare towers over Elizabethan England, or Beethoven over nineteenth-century Vienna. His life's work stands alone, untouched—a major creative act.

When discussing black holes, we are talking about gravitation itself. But what is gravity? Gravity is why things stick to the Earth. Right now everything is standing right side up. Six hours from now the Earth will have spun one-quarter of the way about in its daily rotation, and things will all be sideways. The chair, the table, the glass of water on the table: all will lie at a crazy tilt. Water from the dripping faucet will fall horizontally into the sink. In another six hours everything will be completely upside down. Things will fall up. This magic is accomplished by gravitation.

It was Sir Isaac Newton who invented the modern concept of gravity, and to him it was a force. Things fall because of an attraction between them and the Earth. Our planet exerts a force upon the droplet of water suspended from the dripping faucet: the droplet breaks free and flies toward the Earth. The very Moon itself is not free of this force, and it orbits about us in response.

But the black hole has no place in Newton's scheme of things. It goes beyond Newtonian ideas. The black hole is a child of Einstein's general theory of relativity. General relativity too is a theory of gravitation, but it treats gravity very differently from the way Newton did. To Einstein gravity is not a force at all but a distortion of the very nature of space and time, and its effects are far more subtle than Newton could have realized. It took Einstein ten years to create this theory, and even today it is one of the most fascinating and complex of all the fields of physics.

As with every other physical theory, general relativity's content is summarized in a set of equations—Einstein's gravitational field equations. These equations describe the gravitational field produced by any body, but as with all equations they constitute not so much an answer as a problem to be solved. And this, in turn, is no easy task. In terms of pure mathematical difficulty, Einstein's equations are among the most inaccessible and opaque of all the equations of physics. They are monstrously forbidding and complex. There is not one but sixteen separate equations to be solved, each one a nonlinear partial differential equation for sixteen separate unknown functions. These functions, in turn, have a very subtle interpretation, and even when the equations have been solved there is still much work to be done in arriving at a final understanding. Even today, more than half a century after Einstein formulated his equations, we know very little about their solutions, and there is an entire branch of physics devoted to their study. Scientists who work in this field tend to be a highly mathematical lot, and they think in terms of concepts so abstract as to lie beyond the concerns of even their fellow physicists. They call themselves relativists.

The first person to find an exact solution to these equations was not Einstein himself. He had attempted to do so in his original paper setting forth the theory of general relativity, but he had not gotten very far, and he contented himself there with an approximate solution. It was the German astronomer Karl Schwarzschild who obtained the first rigorously correct solution to the gravitational field equations, and the remarkable thing about his achievement is the manner in which it came about. Schwarzschild did not find his solution in a book-lined study. He found it while at war.

At the outbreak of World War I, Karl Schwarzschild was forty years old, and he was one of Germany's most eminent astronomers. Behind him lay an impressive list of scientific achievements, and he had reached an age at which one might reasonably expect to settle down. Nevertheless his patriotic ideals made him feel it necessary to volunteer for the army. He served first in Belgium and then in France, and then was transferred to the Eastern Front. While in Russia he contracted the rare, painful, and incurable disease of pemphigus. Under the stress of war, under the stress of illness, Schwarzschild continued doing science. He worked out his solution to the field equations. Two months later his illness had become so grave that he was sent home to Germany. Two months later he died.

Schwarzschild's paper announcing his discovery was published in the 1916 edition of the *Journal of the Royal Prussian Academy of Sciences*. It is entitled "On the Field of Gravity of a Point Mass in the Theory of Einstein," and it sits right next to one by Helmreich on "Manuscript Emendations in Galen's Glossary to Hippocrates." The 1916 edition makes a hefty load—it runs to 1,400 pages—and in it are to be found articles on archaeology ("Contribution to the Study of Egyptian Religion"), astronomy ("On the Period of the Variable Star RR Lyrae"), and literature ("On the Upanishads"). There are minutes of Academy meetings and progress reports on its activities. And there is an obituary of Schwarzschild. It was written by Einstein.

Schwarzschild appears to have been pleased with his solution to Einstein's equations. In his paper he comments, "It is always pleasing to have exact solutions to problems," and, somewhat later, that it "permits Mr. Einstein's work to shine with increased purity." But there is no evidence that he took it very seriously. Beyond a certain satisfaction in good work well done, his paper conveys no particular tone whatsoever. Indeed, Schwarzschild's paper is remarkably brief— a mere few pages—and it is almost entirely devoted to mathematics. It sets forth his method of solving Einstein's equations and that is that. Nowhere in the paper is there to be found the slightest indication that something monumental has been found. Nor is there an indication that either Einstein or anybody else believed so. When Schwarzschild's paper was published it aroused no stir. The scientific community simply took note of it and continued about its business.

It is a measure of the complexity and subtlety of general relativity that this could be so. After all, in most branches of science, the hard work is over once a problem has been solved exactly. But here the work had only begun. It took Schwarzschild a matter of months to find his solution. It has taken us half a century to appreciate its full significance. Mathematically speaking, the Schwarzschild solution looks remarkably simple, and it is easy to write down. It is presented in all the textbooks. But beneath this apparent simplicity lies an extraordinary richness. Furthermore, the solution is difficult to understand. It has a number of features that look positively impossible— pathological. So strange do these pathologies appear that physicists did not know how to proceed. As late as 1960 there was serious debate as to how they should be treated. One approach was to ignore the difficulties altogether. A widely respected textbook of general relativity published in 1965 does just this. It hardly mentions their existence. Another approach was to take them more seriously and claim that they threw the entire Schwarzschild solution into question. Einstein himself for a time took this approach, and at one point he published a paper attempting to show that the solution could never pertain to reality. As it turned out, he was wrong.

It has been only recently that the smoke has lifted, and the truly revolutionary nature of Schwarzschild's result uncovered. Far from being mere annoyances, these pathologies are now recognized to be utterly fundamental in their significance. Only now do we understand the true nature of the Schwarzschild solution. The Schwarzschild solution describes the black hole.

Schwarzschild had not been looking for just any solution to Einstein's equations. He had wanted to answer a very definite question. He wanted to study the *gravitational field outside a spherical body*. The Schwarzschild solution was his answer.

The Earth is very nearly spherical, so the Schwarzschild solution applies to it. The same is true of the Sun. But neither of these bodies is a black hole. Their gravitational fields are entirely unremarkable and show none of the bizarre behavior characteristic of black holes. Nevertheless, in some important way the black hole and such prosaic bodies as the Earth and Sun are related. What is this relation?

The relationship is *compression*. Any object can be transformed into a black hole by the simple act of shrinking it. This is the meaning of the imaginary experiment of crushing the Sun conducted above. At every stage in that process the gravitational field of the Sun was given by the Schwarzschild solution. When the Sun was relatively large that field was unremarkable, but once it had been compressed to a radius of 1.75 miles, a dramatic change had occurred: a change in which the Sun turned into a black hole.

This critical radius of 1.75 miles is of vital importance. It is so important that it has a name: *the Schwarzschild radius of the Sun*. It is the radius to which the Sun must be compressed in order for escape velocity from its surface to equal the velocity of light.

As for the Sun, relativists speak of the Schwarzschild radius of any object. The Schwarzschild radius of the Earth is one-third of an inch, that of our galaxy .03 of a light year. Scientists go further and speak of an object's *Schwarzschild surface*. This is the surface of an imaginary sphere whose radius is just the Schwarzschild radius of that object. The Earth's Schwarzschild surface, a mere fraction of an inch across, lies buried deep within it (Figure 26).

The Schwarzschild surface has no material reality. If we were to tunnel into the Earth until we reached a point one-third of an inch from its center, we would find nothing unusual there. It would be like standing in a cornfield precisely on the border between the United States and Canada.

But this is because the Earth is much larger than its Schwarzschild radius. So is the Sun and every star visible in the sky. Indeed, it is difficult to imagine an object so violently compressed that it lies within its Schwarzschild surface. On the other hand, pulsars are almost so small. Certainly nothing prevents us from considering one as in Figure 27.

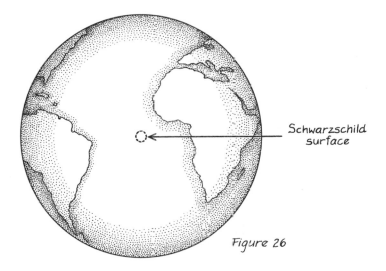

Schwarzschild
surface

Figure 26

In this case the Schwarzschild surface lies outside the object. As in the previous example, it has no material reality. But unlike that example, it now has a vital significance. It is a black hole.

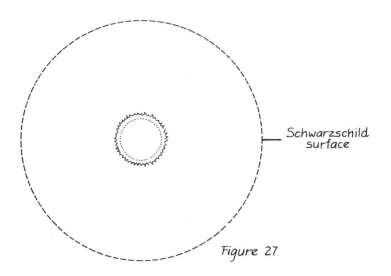

Schwarzschild
surface

Figure 27

9

Light

The scene is the island of Principe, a tiny speck of land washed by the Atlantic Ocean. A few miles to the south lies the equator, and not far over the horizon to the east the African coast. The heat and the humidity are intense. Clouds have gathered, and through them the Sun is dimly visible. It is crescent-shaped. The date: May 29, 1919. An eclipse of the Sun is about to begin.

Eddington and Cottingham have no time to gaze at the spectacle overhead. It is their equipment that concerns them: an eleven-foot-long telescope lying flat upon a table, a mirror mounted before it reflecting the image of the Sun onto its lens, a motor to rotate the mirror as it tracks the moving Sun. Their hearts are heavy as the clouds obscure the sky. They have been a full month on this island preparing for the eclipse, and before that two months more journeying there from England.

As the moment of totality arrives they begin their work. It is performed in silence, with the utmost care, and with the greatest possible urgency. They have rehearsed their motions time and time again in preparation, as a ballet dancer would rehearse a particularly exacting part. One man is inserting a series of sixteen photographic plates into the telescope in rapid succession. The other is exposing these plates for varying lengths of time—two seconds for the shortest, 20 seconds for the longest. The only sound is the steady beat of a metronome ticking out the 302 seconds of the eclipse. Shining over their heads, dimly visible through the clouds, is the unearthly silver radiance of the solar corona. Arching high above it is a mighty solar prominence extending 100,000 miles out from the Sun. The two are entirely unaware of its existence. Not once do they glance upward at the sky. Only days later, when the plates are developed, will they realize what they have missed.

Two hours earlier and thousands of miles away, the very same scene has already been enacted, this time in the small town of Sobral in northeastern Brazil. There too, in the same hurried mad rush, a series of photographs of the eclipse were taken, and there too, these photographs will be developed and scanned with minute intensity. But it is not the image of the eclipsed Sun that so concerns these astronomers. Nor are they concerned with that glorious prominence. They are looking for stars on their photographs—stars whose light has passed close to the edge of the Sun. They intend to measure the positions of these stars. They intend to measure these positions to an accuracy of 1/1,500 of an inch.

They are weighing light.

The 1919 eclipse expeditions tested a prediction of the general theory of relativity. According to this theory, light is affected by gravitation. Einstein was saying that light falls.

To the modern ear, this might or might not sound surprising, but in the year 1919 it was revolutionary. It ran counter to everything people knew about light, both scientists and the general public alike. For after all, in daily experience it was only matter that was heavy. *Things* had weight: material objects that could be touched. But light was not a thing. Light was insubstantial and evanescent: a gleam on the horizon. How could a glimmer fall?

This commonsense view was strictly in accord with scientific law in 1919. Matter was made out of atoms; light was a wave in the ether. The idea that such waves could be attracted to the Earth lay entirely outside the scientific framework of the day. The ether itself might be so attracted—but the proposition that waves within it would too was as silly as the notion that ripples in the middle of a pond might be attracted to a bush growing on its edge. More than that: it was inconceivable. Nothing that physics had learned of light in the past 200 years gave the slightest inkling that this could be so. If Einstein were right, a scientific edifice two centuries in the building was due for a critical modification.

How can it be that light is affected by gravitation? Einstein's famous equation $E = mc^2$ seems to offer an answer. According to this equation, every mass m has associated with it an energy equal to m times the speed of light squared. This energy is the dreadful power of the hydrogen bomb. But the opposite interpretation is also true: to every energy E there is associated a mass equal to E divided by c^2.

According to this view, pure insubstantial light carries mass. That lamp by the bedside is not just emitting radiance—it is emitting something that has weight. This weight is very small. If the lamp were to be left burning for a year, it would have put out a mere one-millionth of an ounce in the form of light waves. Another way of saying the same thing is that at present rates electric light costs a staggering $50,000,000 an ounce—a sum testimony to the minuteness of the mass associated with energy. But although this mass of light may be small, it is still there according to Einstein—and any mass, no matter how tiny, must fall.

Unfortunately this attempt at an explanation will not work. It makes sense but it is simply wrong. It is wrong, in fact, by exactly a factor of two, for light turns out to fall at just twice the rate predicted by this line of thought. A proper understanding of the phenomenon is far more delicate and involves an analysis of the very nature of space and time themselves. We will return to it in Chapter 11. Put the question aside till then.

At any rate it is easy enough to see what Einstein was predicting. Throw a ball horizontally—precisely so. The ball leaves one's hand moving sideways but it does not continue doing so for long. Gravity bends its path downward, and soon it falls to Earth. The same is true of a bullet from a gun—and, according to relativity, to a beam of light from a flashlight. Einstein's prediction is that light emitted horizontally from the surface of the Earth is deflected downward by 0.00015 seconds of arc—1/24,000,000 of a degree.

So small is this angle of deflection that any experiment designed to test the prediction was impossible in the year 1919. It is still impossible. The idea of the experiment is simple enough: one shines a flashlight at a wall and checks to see if the patch of illumination lies slightly below its expected level. But because the bending of light by gravity on the Earth is so weak, the accuracy required for the experiment is far beyond anything that can be achieved.

Because gravity on the Sun is stronger than on the Earth, Einstein would predict a greater angle of deflection there. An experiment to test this prediction would have a greater chance of success. And although we cannot go to the Sun to perform the test, a minor variation suggests itself. Do not fire a light ray off from the Sun's surface. Fire one *past* the Sun as in Figure 28. Graze its surface with a beam of light. Where does this beam come from? It comes from a distant star. So strong is gravitation on the Sun that the angle of deflection of this light ray is quite large: nearly 2 seconds of arc.

The effect of this deflection is to shift the position of the image of

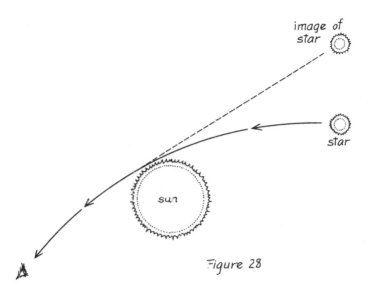

Figure 28

the star. The shift is away from the Sun. Thus, as a result of the bending of light by gravitation, the star acts as if it has been repelled by the Sun—by just the angle of deflection of the light ray. A photograph of stars in the vicinity of the Sun will show that they appear to have moved outward away from their true positions by just this angle.

Such is Einstein's prediction—and this prediction, in contrast to the previous one, is not impossible to test. An angle of 2 seconds of arc may seem small in day-to-day terms, but it is not small to astronomers. Astronomers are used to dealing with such angles. The experiment seems entirely feasible.

There is a rub, of course. Stars cannot be seen in daylight.

Luckily for Einstein—luckily for us all—there is a way out. By a truly remarkable coincidence the Moon appears to be just the size of the Sun. It is an illusion of perspective, of course. In reality the Moon is far smaller than the Sun. But it is also much closer to us— just exactly the right amount closer. Had the Earth no Moon at all, or had our Moon been smaller or farther away, there would be no such thing as a total eclipse of the Sun. Had our Moon been bigger, or closer, a total eclipse would be more like an ordinary nightfall. Only if the Moon and the Sun just happen to have the same apparent size in the sky will an eclipse be the spectacle we observe it to be. *None* of the other planets of the Solar System has such eclipses.

But we do. Every few years the shadow of the Moon falls somewhere on the face of the Earth. Go and stand there and you will witness one of the most extraordinary of all cosmic spectacles. The blaze of light from the Sun is extinguished. Night falls. The corona of the Sun—forever shining, almost never visible—now blazes forth in a magical pearly glow. If you are doubly lucky a mighty prominence will hover like a motionless cosmic flame. The sight is enough to bring tears to the eyes of the most hardened of astronomers.

And the stars come out.

Perhaps it is best to continue now in the words of Sir Arthur Eddington, one of the foremost astronomers of his day and the leader of the two eclipse expeditions.

> *The bending affects stars seen near the Sun, and accordingly the only chance of making the observation is during a total eclipse when the Moon cuts off the dazzling light. Even then there is a great deal of light from the Sun's corona which stretches far above the disc. It is thus necessary to have rather bright stars near the Sun, which will not be lost in the glare of the corona. Further, the displacements of these stars can only be measured relative to other stars, preferably more distant from the Sun and less displaced; we need therefore a reasonable number of outer bright stars to serve as reference points.*
>
> *In a superstitious age a natural philosopher wishing to perform an important experiment would consult an astrologer to ascertain an auspicious moment for the trial. With better reason, an astronomer today consulting the stars would announce that the most favorable day of the year for weighing light is May 29. The reason is that the Sun in its annual journey round the ecliptic goes through fields of stars of varying richness, but on May 29 it is in the midst of a quite exceptional patch of bright stars—part of Hyades—by far the best star field encountered. Now if this problem had been put forward at some other period of history, it might have been necessary to wait some thousands of years for a total eclipse of the Sun to happen on the lucky date. But by strange good fortune an eclipse did happen on May 29, 1919.*

Stars by and large are unchanging, and astronomers usually have plenty of time to conduct their observations. If the equipment does not work this night, one can always try again on the next. Not so with this observation, however. Eclipses of the Sun are few and far between, and when they do come they are exceedingly brief—a matter of minutes at most. Into these few short minutes the entire observation must be compressed, and if it fails there is no way of making up the loss short of waiting till the next eclipse. Not only that, but as Eddington has emphasized, not all eclipses are suitable for weighing light. If anything had gone wrong, a full nineteen years would have passed until the next opportunity. So stringent was the accuracy required that the most subtle errors could have ruined everything—as in fact occurred with some of the data obtained at Sobral. Astronomers observing an eclipse have also been known to blunder into their equipment and send it toppling to the ground in their urgency. Finally, so simple a matter as a heavy cloud cover could have scotched the attempt.

Accordingly, not one but two expeditions were sent out, and each was planned with scrupulous care. It was the Astronomer Royal, Sir Frank Dyson, who had realized two years before the rare opportunity for testing general relativity provided by this eclipse, and to him much of the credit for the expeditions' success is due. And it was through Dyson's efforts that Eddington was chosen to direct them.

This did not exactly come about in the usual way. World War I was at its height—and Eddington was a Quaker. He was a deeply religious man, and if drafted he was resolved to declare as a conscientious objector. Nowadays conscientious objectors are looked upon as men of high moral stature, but in 1917 they were pariahs. So fervent was the patriotic climate of the day that many of Eddington's colleagues at Cambridge felt their university would be disgraced if he were to declare. Accordingly, the Old Boy network went into operation. Appeals were made to the Home Office to defer him from the draft on the grounds that it was in the national interest to keep so distinguished a scientist at home. The Home Office was willing to cooperate, and for a time it appeared that the attempt would succeed.

It was Eddington himself who threw the monkey wrench into the works. The Home Office mailed a deferral form to him. All he had to do was sign and return it. But at this very moment friends of his were languishing in internment camps for expressing the very same convictions as his. Eddington decided that he could not allow himself to go free on so shabby a pretext. So he returned the form unsigned, but added a note that if he were not deferred he would claim conscientious objector status anyway.

This short note triggered an uproar. By the letter of the law, the

Home Office now had no choice but to send him to an internment camp. Far from being solicitous, many of Eddington's colleagues at Cambridge were furious with him. They felt he was only striking a high moral stance, and that by it he had precipitated a crisis of his own making. Eddington in turn expressed no anxiety at the prospect of imprisonment, and claimed surprise at their annoyance. He was simply living up to his principles.

Into this pickle stepped the Astronomer Royal with a saving proposal. Dyson suggested that Eddington be deferred on the express stipulation that he plan and lead the eclipse expeditions of 1919. It was an ingenious compromise, and one that was acceptable to everyone. And so Sir Arthur Eddington became the first man to test experimentally the general theory of relativity.

It may be difficult to recall now the magnitude of the hatred of Germany aroused by the war. In America it was widely forbidden to teach the German language in public schools. In England the royal family changed its name from the House of Hanover—a German name—to Windsor. Numerous British scientists had died on the battlefield, and there was widespread anti-German sentiment in the universities. It was even a serious question whether scientists should have anything to do with their German colleagues, and not just during the war but long after. Relativity, the product of a German physicist, was not immune from this attitude. Relativity was enemy science.

Not only that, but Einstein's recent work was largely unknown in England at the time. The special theory of relativity, published in 1905, had seeped across the Channel by then; but the general theory, which came out during the war, was another matter. In particular, Einstein's prediction of the bending of starlight by the Sun was hardly common knowledge. Scientific communication between the two sides had come to an almost complete standstill by the last years of the war. German and British scientists had ceased exchanging letters. Subscriptions to enemy scientific journals were all canceled. Holland, on the other hand, being a neutral country, did receive these journals, and one Dutch astronomer regularly forwarded copies of Einstein's papers to Eddington out of personal friendship. Through this fragile pipeline Einstein's work filtered into England. Eddington's, in fact, were the *only* copies of Einstein's papers available to the British in those days, and it was largely he who was responsible for bringing relativity to their attention. It was through his friendship with Eddington that Dyson, the Astronomer Royal, had first heard of the deflection of light by the Sun.

All in all, it is very much to the credit of British science that the 1919 eclipse expeditions ever came to pass. They were planned in the teeth of an intense hatred of everything German, and in the midst of a widespread ignorance of Einstein's work. They were also put together at the very last minute. No work could be done on fabricating the equipment required for the observations while the war was in progress. Not until the armistice was signed in November of 1918 could the instrument makers begin their task. It was completed in a rush, and three short months later the expeditions set sail. Eddington and E. T. Cottingham went to Principe; two others went to Sobral.

There were three possibilities. On the one hand, the observations might reveal no distortion in the positions of the background stars at all, in which case light would not be subject to gravitation. Or there might be a "half-deflection" of the sort described earlier in this chapter. Finally, there might be the full two-seconds-of-arc deflection predicted by general relativity. "I remember Dyson explaining all this to my companion Cottingham," Eddington later wrote, "who gathered the main idea that the bigger the result, the more exciting it would be. 'What will it mean if we get double the deflection?' 'Then,' said Dyson, 'Eddington will go mad, and you will have to come home alone.'"

Eddington and Cottingham were two months at sea, and another month on Principe completing their preparations. The day of the eclipse dawned cloudy, and many of their photographs were ruined by the obscuration. But on some the stars shone through. The best showed images of five stars, and it was measured on the spot. "Three days after the eclipse, as the last lines of the calculation were reached, I knew that Einstein's theory had stood the test," Eddington wrote. "Cottingham did not have to go home alone."

It was not until months had passed that any further confirmation was obtained. Four photographic plates were brought back to England undeveloped, since they were of a sort that could not be developed in hot weather. Of these one showed good stellar images, and upon measurement it too showed the deflection predicted by general relativity. The Sobral party remained in Brazil for two months after the eclipse in order to get good comparison photographs of the star field after the Sun had moved away from it in its journey across the Zodiac. When they returned, their photographs were found to contain by far the best data—they had had good weather for the eclipse. It was these that ultimately provided the most accurate test.

THE GLORIOUS DEAD
ARMISTICE DAY OBSERVANCE
ALL TRAINS IN THE COUNTRY TO STOP

The King invites all his people to join him in a special
celebration of the anniversary of the cessation of war, as set
forth in the following message:
"It is my desire and hope that at the hour when the armis-
tice came into force, there may be, for the brief space of two
minutes, a complete suspension of all our normal activities.
During that time . . . all work, all sound, all locomotion
should cease, so that, in perfect stillness, the thoughts of
everyone may be concentrated on reverent remembrance of
the Glorious Dead."

This article was carried in the November 7, 1919, edition of the
London *Times*. There is another article on the very same page:

REVOLUTION IN SCIENCE
NEW THEORY OF THE UNIVERSE
NEWTONIAN IDEAS OVERTHROWN

Yesterday afternoon in the rooms of the Royal Society, at
a joint session of the Royal and Astronomical Societies, the
results obtained by British observers of the total solar eclipse
of May 29 were discussed.
The greatest possible interest had been aroused in scien-
tific circles by the hope that rival theories of a fundamental
physical problem would be put to the test, and there was a
very large attendance of astronomers and physicists. It was
generally accepted that the observations were decisive in
verifying the prediction of the famous physicist Einstein. . . .

Alfred North Whitehead was there:

The whole atmosphere of tense interest was exactly that of
the Greek drama: we were the chorus commenting on the
decree of destiny as disclosed in the development of a su-
preme incident. There was dramatic quality in the very
staging—the traditional ceremonial, and in the background
the picture of Newton to remind us that the greatest of
scientific generalizations was now, after more than two

centuries, to receive its first modification. Nor was the personal interest wanting: a great adventure in thought had at length come safe to shore.

The *Times* article concluded with the comment that "even the President of the Royal Society, in stating that they had just listened to 'one of the most momentous, if not the most momentous, pronouncement of human thought,' had to confess that no one had yet succeeded in stating in clear language what the theory of Einstein really was." And it then went on to state, in one exceedingly short paragraph, what the theory of Einstein really is.

A third article on the same page announced that no private person could own, store, or drive a motor car in all of Ireland without the express permission of the government. The weather prediction for the day was "dull and cold: some showers or drizzle."

The observation of the bending of starlight by gravity aroused a storm of public excitement. Hardly ever has a scientific discovery received so much attention from the press. The prediction of a German physicist had been verified by British astronomers in an expedition to distant lands. It was something that could not help but appeal to a world reeling from the most devastating war in history. It also helped that Einstein was a pacifist, and had spoken out often and at great personal risk against the war. But most important of all was the widespread public recognition that Einstein had made this prediction on the basis of a truly revolutionary line of argument. The public knew that he was striking at our most universally held ideas of space and time. The eclipse expeditions of 1919 mark the beginning of Einstein's elevation to his present nearly mythical stature. Before 1919 he was famous to physicists. After 1919 he was famous the world over.

At any rate he seems to have been pleased by the event, if we can judge from a postcard he wrote that fall:

Dear Mother,
   Good news today. H. A. Lorentz has wired me that British expeditions have actually proved the light deflection near the Sun. . . .

Leave now the bending of light by the Sun and pass on to the analogous bending by black holes. The angle of deflection produced by the Sun is small only because gravity on the Sun is weak in astronomical terms. In contrast, the black hole's gravitational field is far more intense and produces effects that are spectacular. Plumb these effects with the aid of a flashlight. Suspend this flashlight over the hole and trace the path of its beam.

Begin by suspending it high above the hole, as in Figure 29. At such great distances the gravitational field is very weak, and the bending is minute: There is nothing new here. But now lower the flashlight (Figure 30). The gravitational field of the hole grows more intense: the path of the light ray more severly curved. This is something entirely beyond our daily experience. Living in such an environment would be like living in a hall of trick mirrors in an amusement park. Objects that appeared directly in front of one would lie below in reality. Objects directly in front would appear above. If a car were driving directly toward an observer, it would seem to be floating high above his head—and the observer would see, not the car's radiator grill, but its roof. The intense gravity from the star below radically distorts the propagation of light and an observer might feel that he was living within a lens. He would be right.

Figure 29

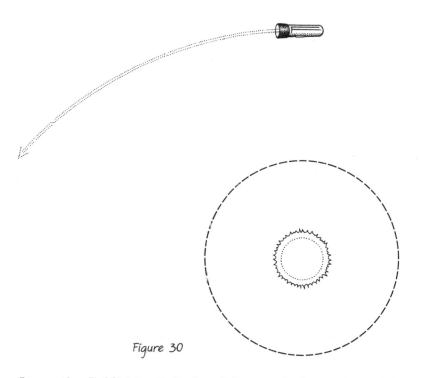

Figure 30

Lower the flashlight still farther (Figure 31). Lower it until its distance from the star is just 1½ Schwarzschild radii. It is now half a Schwarzschild radius from the Schwarzschild surface, and gravity has grown so strong that the path of the light ray is bent into a circle. The beam of light orbits the star.

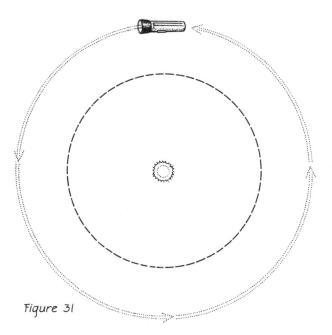

Figure 31

Life in such an environment would offer truly remarkable sights. An observer would see the back of his head hovering directly in front of him. Off to one side would be his ear. Surrounding him in a ring would be images of his head, all viewed from different angles. From this position we can also tilt the flashlight upward (Figure 32). In

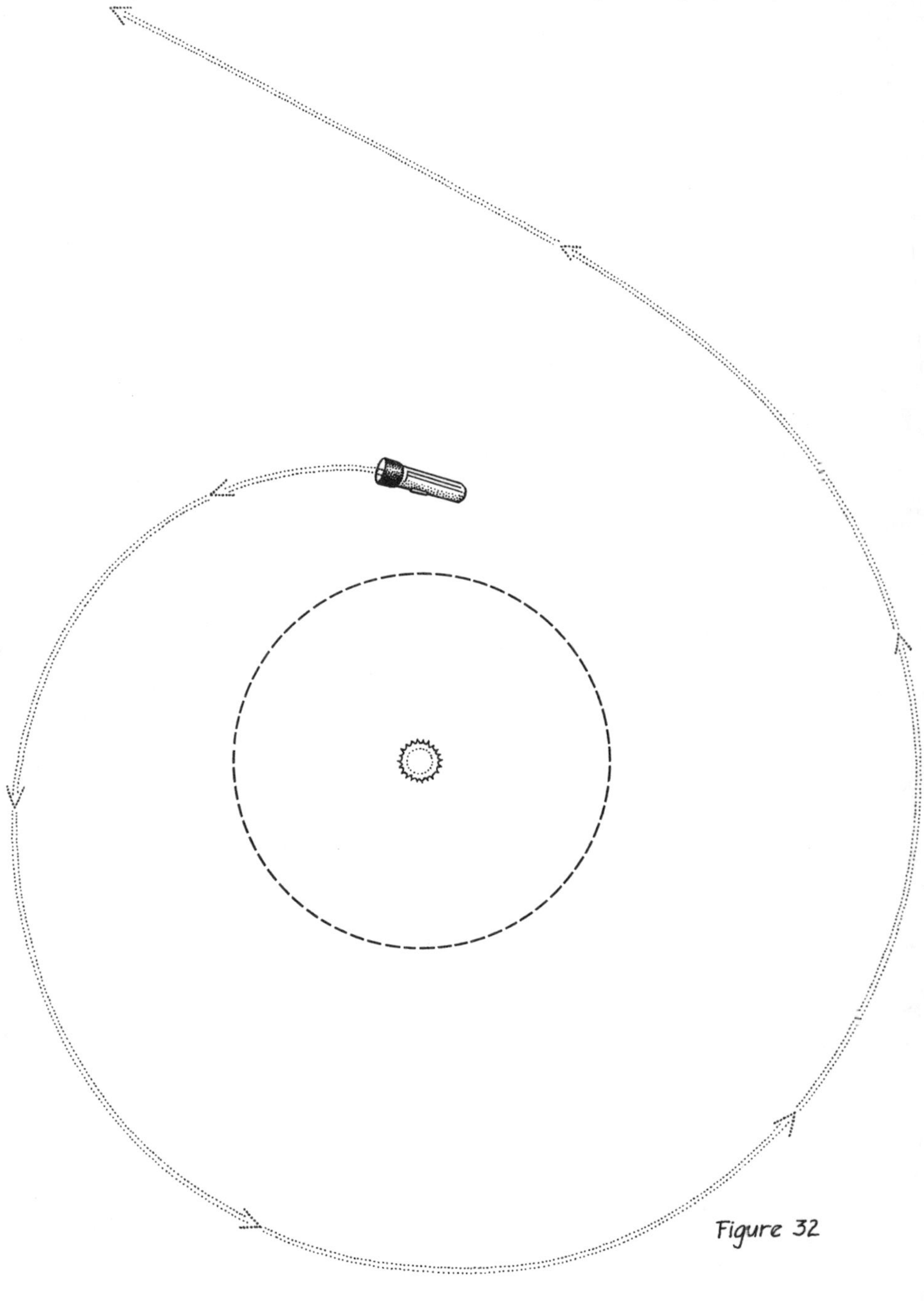

Figure 32

this case the beam slowly spirals outward, ultimately propagating into regions in which gravity is sufficiently weak that its path is essentially straight. If we tilt the flashlight down (Figure 33), the beam spirals

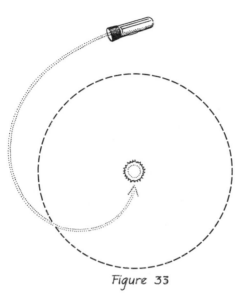

Figure 33

inward, eventually falling upon the star below. This light has been trapped by gravitation.

The closer the flashlight approaches the star, the more of its light is trapped. Once it has been lowered onto the Schwarzschild surface, *all* of its rays are sucked downward and absorbed. Even if we angle the flashlight upward here, gravitation is so strong that its beam arcs onto the star. The same is true (Figure 34) if the flashlight is lowered

Figure 34

through the Schwarzschild surface. No ray of light emitted from within this region is capable of escaping to the distant stars. Light falls like a stone.

Now do a different experiment. Lower not a flashlight but a light bulb toward the Schwarzschild surface. Hover far out in space with a long coil of electric cord. At the end of this cord shines the bulb, dangling beneath us above the star. Pay out the cord and *look* at the light bulb.

As the bulb drops down it begins to grow abnormally dim. It is as if the electricity in the cord were failing. It is not failing, though. In reality the bulb shines as brightly as ever—but as it is lowered less and less of its light reaches us. The light is being sucked down onto the star. The closer the bulb approaches the Schwarzschild surface, the dimmer it grows, and once it reaches this surface it has dimmed to invisibility. The bulb has passed from sight.

An object lying within the Schwarzschild surface cannot be seen. No matter how brightly it shines, its light never reaches us. This is not only confined to light bulbs. It is true of everything. Drop a brick onto the star. One sees this brick by means of the light it reflects, but once it has crossed that magical surface, all of this light is bent down and away from us. The brick vanishes the moment it crosses the Schwarzschild surface.

Not just light is affected in this way. Radio signals, like light, are waves in the electromagnetic field, and they too are attracted by gravity. If a radio transmitter were to be lowered, its transmissions would appear to die out just as it passed from sight.

The same is true of X rays and gamma rays as well. The Schwarzschild surface is capable of containing the most intense levels of radioactivity. Nuclear radioactivity among other things consists of these rays—ultrahigh-frequency electromagnetic waves. (It also consists of high-energy particles but as shown in Chapter 10, these too are trapped.) The most dangerously radioactive substances can be rendered harmless in this way. Pour a truckload of waste uranium from a nuclear reactor onto a highly compressed star and then hold a Geiger counter close to its Schwarzschild surface. Inches away lie tons of deadly uranium, but not a shred of this radioactivity penetrates. The Geiger counter is silent.

What of the star itself? All these effects are products of its gravitational field, but it is not even visible. Just as light from the bulb is sucked downward, just as radio waves and gamma rays are contained, so too is the light of the star trapped by gravity. It never even rises above the star's surface, so strong is its pull. *An object smaller than its Schwarzschild radius cannot be seen.* Nor would it show up on a radar screen. Radar works by bouncing radio signals off things: the return echo is detected and signals the presence of the reflecting object. But this star absorbs.

The impenetrability of the Schwarzschild surface has led physicists to coin for it another name as well: *the horizon.* It is an apt description. Things beyond the horizon are there all right, but they are invisible. The difference is that this horizon is absolute. On Earth the horizon retreats if we walk toward it, but the Schwarzschild horizon is inviolate. Once hidden within it, an object is lost from sight forever.

The horizon is the surface of the black hole. A black hole is a region of space lying within a certain critical distance of a highly compressed star. The more one studies this region, the more the star responsible for it recedes into the background, and the more the hole itself—mere, insubstantial space: a collection of places—assumes an almost material reality. It is spherical and it has a definite size: the Schwarzschild radius of the star within. It is black and neither emits nor reflects light. It has a mass: the mass of the star that made it. And it has a number of other strange properties as well.

We close this chapter with a brief excursion into a remarkable coincidence. All this flows from Karl Schwarzschild's solution to Einstein's gravitational field equations—but Schwarzschild's name means something. It is composed of two German words: *schwarz,* meaning black; and *schild,* meaning a barrier, a shield. And what is the black hole's horizon but a black shield hiding what lies within?

Schwarzschild's name describes what he discovered.

10

Force

What is the gravitational field of the black hole? With what force does it attract things to it? On the Earth we answer such questions every day. We do this by picking things up. Every time I lift a suitcase I have measured—or at least experienced—the force with which the Earth attracts that suitcase. Do the same with the black hole. Suspend some object over the hole at the end of a rope, and measure the force on the rope.

To make things specific, return to the previous experiment of lowering a light bulb into the black hole. Chapter 9 asked what the bulb looked like. Now the question is how much it weighs. Newton teaches—and Einstein agrees—that this attraction increases as the light bulb approaches the source of the gravitational field. Begin very far away: 10,000 miles from the hole.

Because the black hole is so very massive, the attraction is very great. Even here the light bulb weighs a full 3,000 pounds. Now pay out the rope. When the bulb is lowered to an altitude of 4,000 miles above the invisible star, the attraction is so strong that it weighs a full 20,000 pounds. At an altitude of 100 miles an incredible 17,000 tons is the force required to keep the bulb from falling. By now it is beginning to dim. The closer it approaches the edge of the black hole, the fainter it grows—and the more it weighs. Its weight grows to unthinkable proportions. Finally, just as the bulb touches the Schwarzschild surface, two things happen. The bulb dims to invisibility—and the force on the rope becomes infinitely great.

The rope snaps, and the bulb falls into the black hole.

Within a black hole gravitation is overwhelming. It is irresistible. An object inside the horizon does not weigh a million tons, or a hundred billion tons. Its weight has passed beyond all numbering and no countervailing force, no matter how enormous, has the slightest chance of supporting it. Once something has touched the edge of a hole its fate is sealed: it falls headlong, it falls helplessly, it falls at the speed of light, and it crashes onto the surface of the star within a fraction of a second. Haul on this object with a rope, seeking to arrest its fall. The rope is torn from your grasp, and it too snakes into the hole. Enter the hole on board a rocket, and fire its engines in an attempt to escape. No matter how hard they are fired they are not being fired hard enough. The rocket falls.

Gravitation inside a black hole has a completely different character than it does outside. In normal circumstances gravity is certainly important, but it is something that can be dealt with. It is a problem that can be solved. Our legs can support us: buildings can be raised. But within the Schwarzschild surface none of these things is even conceivable. Once within this surface no one would be able to stand. Buildings would fall. Light itself would fall. Inside the black hole, falling is a way of life.

This fall is shared by everything, and it goes on for a certain period of time. It goes on until the falling object has struck the star below. As for how long this takes, it depends on the size of the hole—and this depends on the black hole's mass. If the star producing the hole has a mass just that of the Sun, its Schwarzschild surface is fairly small—1.75 miles in radius—and objects fall for only 1/100,000 second before dashing themselves to bits. But larger black holes may also exist. If somewhere in the universe an entire galaxy has been compressed within its Schwarzschild radius, the black hole it has become would be hundreds of billions of miles across. Objects crossing within would enter into the state of helpless plunge for a period of ten days. It is even conceivable that there may exist still larger holes—holes so massive and so vast that things within fall for centuries. Would life be possible in such an environment? Perhaps it could, and if so, it would be a strange way of life indeed. The reader may find it amusing to speculate on what sort of world this would be in which to live: the world of the Universal Fall.

The black hole devours everything it touches. Suppose that a hole 1 foot across has entered the room and floats before me. Such a hole would be produced by an object a good deal less massive than the Sun, but big enough for all of that: it is about 18 times as massive as the Earth. I cannot see it. All I see is a 1-foot disc, utterly black, silhouetted against the background wall; and around this silhouette the image of the wall, itself radically distorted.

In my hand is a yardstick. Gingerly I reach forward and touch its tip against the hole. The yardstick is snatched from my hands and vanishes in a trace. The hole is 1 foot across, the yardstick 3 . . . and the yardstick is inside the hole. It was crushed to fit inside.

I shove a sofa toward the hovering darkness. It crumbles like a sheet of paper and disappears from sight. Beside me is a heap of malodorous garbage—orange peels, coffee grounds. In they go. Even the smell is absorbed.

I throw a hand grenade down the hole. No sound of an explosion reaches my ears; no deadly hail of shrapnel cuts through me. The hole does not even quiver as the grenade goes off inside. It has contained the explosion. If it had been a hydrogen bomb I had thrown in, the hole would have contained that too.

There is a further element in the situation. What is the force this hole exerts on me? I have been giving it a wide berth. I have no wish to be dashed to bits on some invisible planet 18 times as massive as the Earth so I have avoided touching its edges. But it does attract me, and although the attraction may not be infinite it is certainly large. How large?

The force of attraction works out to some 10,000,000,000,000,000 pounds.

I am catastrophically jerked off my feet and plunge headlong into the hole. Desperately I grab at a doorknob in my flight but it is ripped from my hands. The table beside me also plummets sideways. The door is torn from its hinges and follows a moment later. A continual roar, an earsplitting shriek, comes from the window as the very air, the atmosphere of the Earth, pours through in a torrent, sucked violently inward. The air around the edges of the hole grows white hot as it is heated by friction and compression. The ceiling crumbles down upon me. The walls of the room cave in. The floor rises up, splintering, pulverized, a geyser of masonry. The tree outside the window bends and snaps like a twig. It shoots violently sideways, smashes against bricks and wood, and all shoulder their way into oblivion.

The hole is hovering over the East Coast of the United States. Three thousand miles away to the west the Pacific Ocean laps the California shore. It too is attracted into the hole, and it rises from its bed in the greatest tidal wave of all time. Los Angeles is inundated as the ocean pours east—not just some of it, not just a wave, but every last drop of water in the entire Pacific Ocean. The waters

I   The Crab Nebula
California Institute of Technology and the Carnegie Institution of Washington.

II   The Pulsar Observatory in the Quabbin Reservoir.
Mr. George Crsten, Department of Astronomy, University of Massachusetts.

III   A Negative photograph of the galaxy Messier 87
*Courtesy Dr. H. C. Arp, Mount Wilson Observatory and Las Campanas
Observatories.*

IV   The Andromeda Nebula, a galaxy similar to our own.
*California Institute of Technology and Carnegie Institution of Washing-
ton, Copyright 1959.*

A computer-enhanced photo
of the jet.

V   The M87 jet.
*Courtesy H. C. Arp, Mount Wilson and
Las Campanas Observatories.*

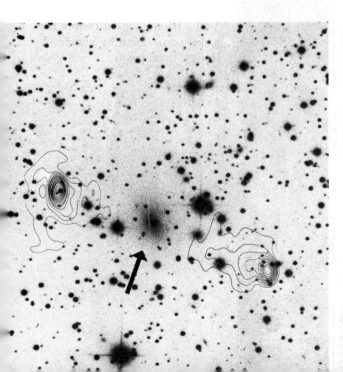

VI   The Galaxy Cygnus A
(indicated by arrow) and its
associated radio-emitting
lobes (negative photograph).
*Courtesy H. C. Arp, Mount
Wilson and Las Campanas
Observatories.*

VII   The Quasar 3C48.
*Copyright 1972 McGraw-Hill, Inc.;
from Hodge, "Slides for Astronomy."*

VIII   Uhuru, tucked away in the
rocket's nose, awaiting launch.
Harvey Tananbaum.

IX   Artist's conception of the Uhuru Satellite in orbit.
Smithsonian Astrophysical Observatory.

X   Artist's conception of a neutron star in orbit about an ordinary star.
Lois Cohen, Griffith Observatory. Courtesy TRW Systems Group.

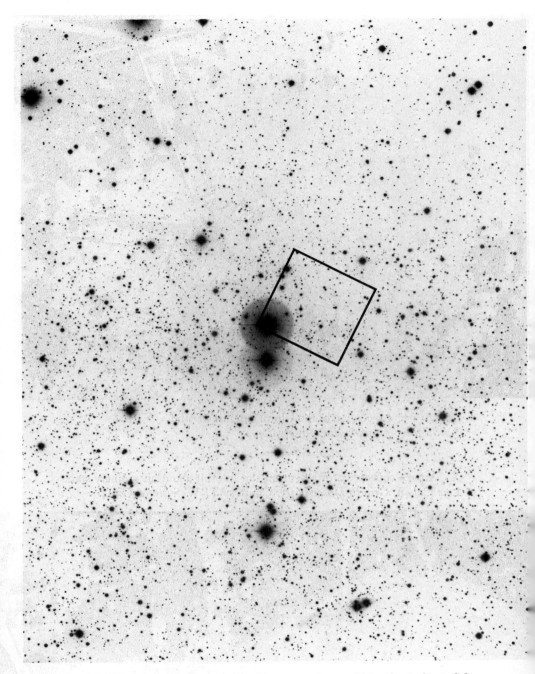

XI   The X-ray source Cygnus X-1 lies somewhere within the indicated box. *Courtesy Dr. Jerome Kristian, Mount Wilson and Las Campanas Observatories; Carnegie Institution of Washington.*

XII   S. Jocelyn Bell Burnell
Courtesy Dr. S. Jocelyn Bell Burnell.

XIII   G. Richard Huguenin
in a pensive mood.
University of Massachusetts.

XIV   Harvey Tananbaum
Karen Tucker

XV   Riccardo Giacconi
Karen Tucker

XVI   Subrahmanyan Chandrasekhar
University of Chicago

XVII   Stephen Hawking
David Montgomery/Syndication Sales

accelerate in their horizontal fall; within a matter of seconds they are thundering along at a hundred miles an hour. Los Angeles is annihilated—but not by the ocean alone. Each person walking the streets has been pulled eastward with a force of several thousand pounds. Skyscrapers topple sideways. Beneath them the very ground itself—rock, dirt, gravel, the fabric of the Earth—all is torn from its bed and catastrophically plunges toward the hole.

The entire planet Earth is absorbed into the black hole. It is folded and crushed, compressed into a ball of debris, and annihilated. The Earth does not fall into the hole all at once, however. The process of destruction takes some time. An avalanche of boulders pouring one way is met by a similar avalanche coming in at an angle. A cosmic traffic jam ensues. This jam slows the pace of events but it cannot halt them. A jumble of debris surrounds the hole. Deep inside a one-foot mouth is surrounded by a blaze of vaporized rock, superheated by the intense compression it has undergone. On the surface of the ball vast geysers, explosions of white-hot vapor, spout from the deep interior. The jumble shrinks. Now it is a mile across. Now it is a yard. Now it is gone. The Earth has been annihilated by a 1-foot area of nothingness.

Finally, only the hole remains. It does not quiver. Only one minor change has taken place. Because the hole has absorbed so much mass, it has grown a little heavier. The Schwarzschild radius is then a little bigger. The mouth grows as it devours.

Matter that encounters a black hole falls into it, and in this process a good deal of energy is released. Just as a collapsing building makes a mighty thump, so the downpour of matter into a black hole has its own characteristic signature. This signature is a violent blaze of radiation—light, radio, and X rays—produced by the enormous heating this matter undergoes as it is compressed to fit into the hole. Once this falling matter has passed over the horizon its radiation cannot be detected—but it also emits long before it has reached this point. Even when it is thousands of miles out from the black hole, the infalling matter is so compressed, so shaken by buffets from every direction, that it radiates fiercely. And this emission is something that can be received.

We have observed such emissions. The sky at night may give the impression of stillness and serenity, but the more astronomers have observed the universe, the more violent a place they have found it to be. It is full of cataclysms. Are these cataclysms evidence of black holes?

The Crab Nebula is the first entry in Charles Messier's catalog of diffuse nebulae. Figure III (photo section) shows another entry—the 87th in his list. Messier 87, M87, as this nebula is known, lies in the constellation of Virgo and it is easily found in the spring and summer sky. Through a small telescope it glows softly with a pearly light. Unlike the Crab it is not a cloud. It is a galaxy—an enormous swarm of stars situated far off into space. The Earth is located in such a galaxy, the Milky Way Galaxy, but it is very different. Figure IV (photo section) shows what *our* Galaxy would look like from a vantage point millions of light years away. From such a distance the Sun would be totally invisible—lost in the swarm: an infinitesimal speck. As for our Galaxy, it is disc-shaped, and it is shot through with winding spiral arms.

M87, on the other hand, has no spiral arms, and it is not disc-shaped but spherical. Also visible in Figure III is a spherical cloud of what appear to be fuzzy dots surrounding it. Hundreds can be counted. These are not stars, but swarms of stars themselves. Upon magnification each is revealed to be itself spherical—miniature versions of the mighty galaxy about which they swarm. They are referred to as globular clusters, and each contains up to 100,000 stars. So vast is the bulk of M87 that they show up as mere smudges beside it.

Galaxies like M87 are common—literally millions are known—and for years this one attracted no special attention. Recently, however, this view has changed. Bit by bit the old picture of M87 as a simple swarm of stars has been eroded away by a series of discoveries. The first was the observation of radio emission from the galaxy. Unlike the pulsars this blaze is steady, unchanging—but it is overwhelmingly more intense. Compared to the magnitude of the radio radiation from M87, the strength of emission from a pulsar is positively meager.

In Figure III the core of M87 appears to be completely packed with stars, a solid mass of light. But this is an illusion. The illusion arises because the core has been overexposed in the photograph, which is a long time exposure designed to reveal the faint outer edges of the galaxy and its accompanying swarm of globular clusters. In Figure V (photo section) is shown the result of a shorter exposure. Here the globular clusters and the outskirts of M87 are quite invisible, and one is penetrating deep into the galaxy's heart. Visible in Figure V is the dense, tightly packed assemblage of stars lying at its very center. But also visible is something else—something that was entirely washed out by the overexposure of Figure III.

This is the jet extending out from the center of the galaxy. It is a remarkable object. In the first place it is *straight*—and straight lines are not often found in nature. In the second place the M87 jet is also a strong source of radio emission. And finally, it is composed of knots.

The knots are more clearly visible in the inset to Figure V, which is an enhanced view made from several photographs joined together in a computer. The M87 jet is not a continuous line at all: it is a series of bunches. The same is true of the radio emission from the jet. And finally, the figure reveals something else of great importance that one might have guessed but could not have known for sure: the jet extends down to the very center of the galaxy.

This center—the *nucleus* of M87—is also a source of radio emission, and X rays as well. But its most important property was not discovered by radio astronomers or X ray astronomers at all. It was made by the simple act of counting stars. This work was done by a team of four astronomers from the California Institute of Technology—Peter Young, James Westphal, Jerome Kristian, and Christopher Wilson—in conjunction with Frederick Landauer from the Jet Propulsion Laboratory. It was spread over a period of more than two years and much of this time was devoted to assembling one of the most sensitive "cameras" that has ever been built: the SIT detector system. The SIT is a Silicon Intensifier Target television camera tube, and it is connected to the eyepiece of a telescope. It replaces the conventional photographic plate and yields results far superior to it. The difficulty is that it produces not a photograph but a complex maze of data that can only be analyzed by a computer. But once this analysis is complete a picture can be assembled, and it has a finer resolution than anything that might be obtained in the ordinary way.

Young and his colleagues turned their device on M87. Their television screen was divided into a multitude of tiny squares, or "pixels" (for "picture elements"): 256 of them on a side. The data they obtained was a series of numbers—256 × 256 of them, each representing the total quantity of light within a given pixel. A startling result emerged when they graphed these numbers. They chose a line of pixels running directly across the image of M87 on the TV screen, and they were careful to make sure that the line ran precisely through the galaxy's center. Figure 35 is a graph of what they found.

There is a strong concentration of light toward the center. This light comes from stars, and the graph shows that they are packed much more tightly in M87's nucleus than at its outskirts. In itself this packing is nothing new: all galaxies exhibit it. It is the degree of packing that is significant here: the nucleus of M87 is *too heavily populated with stars*.

The astronomers convinced themselves of this by examining a second galaxy, one with the prosaic name of NGC 4636. The object was chosen to be as similar to M87 as could be arranged: it too was

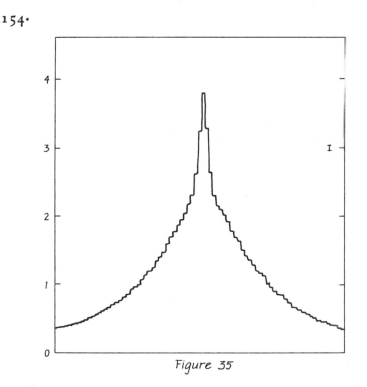

Figure 35

spherical, contained no spiral arms, and was roughly the same distance away. The results are shown in Figure 36. This figure too shows an increase of light as one moves inward, but the increase is nowhere near as dramatic. In particular the "spike" of light at the

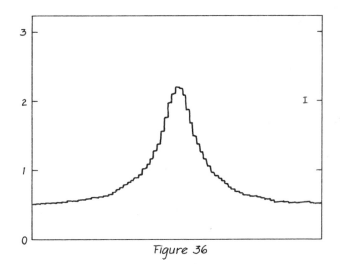

Figure 36

nucleus of M87 is absent in NGC 4636. It is this central spike that is unusual and significant about M87. No other galaxy shows it.

What can be causing so great a packing of stars at the galaxy's center? Something must be holding them there. Some enormous concentration of matter must be pulling them in. On the other hand this concentration itself is not showing up on the SIT data. These data show only the light from stars, and nothing from any additional giant lump at the center. Whatever this lump must be, it cannot be emitting very much light.

Perhaps it is not emitting any light at all. Perhaps it is a black hole.

Their knowledge of the distribution of stars in the nucleus of M87 enabled the group to calculate both the mass and size of this central object. They found its mass to be a gigantic five *billion* times that of the Sun, and its diameter less than 600 light years—possibly very much less. Could this object be a giant black hole? Certainly it is a possible interpretation of their result. And it jibes with other pecularities shown by the galaxy as well. The intense radio and X-ray emission from its nucleus fits in with the idea, since the hole would be expected to swallow continually any matter about it, and in so doing emit radiation.

Whether or not this model of M87 is correct, it is clear that we have a long way to go in understanding the galaxy. Nothing Young and company have found helps to explain the jet extending out from the location of the black hole, nor the many small knots of which it is composed. Nor has it been conclusively shown that M87 really does contain a black hole. That is a supposition, and although it is consistent with their data, it is not the only possible interpretation. *Anything* composed of five billion stars will explain their discovery, provided only that it is smaller than 600 light years in extent and emits small quantities of light. The central object may turn out to be nothing more strange than 5,000,000,000 abnormally dim stars clustered into the required volume of space. And 600 light years is a very big distance indeed.

In order to understand what Young and company found, it is going to be necessary to do them one better. We will have to repeat their observation but with a finer device. If it turns out that the central concentration of stars in M87 is even more dramatic than they had found, the black hole interpretation will be proved. Alternative explanations not involving a black hole all predict a relatively mild spike in this distribution, and they would be ruled out by such a discovery.

The problem is they did the job too well. They did it as well as can be done. With presently existing telescopes and instrumentation there is no possibility of improving upon their work. There seems to be only one prospect on the horizon for ever resolving this question: the space telescope. This will be a large telescope placed in orbit about the Earth and serviced by the space shuttle, and when in full operation it will be far and away the most powerful telescope in the world. It will revolutionize our view of the universe. Among other things it will be ideally suited for deciding whether M87 contains a giant black hole in its center. At the moment the space telescope is scheduled for launch in the late-1980's, but given the budget problems NASA faces it is hard to know if this will really come to pass.

Until it flies we will simply have to wait.

The wispy blob at the center of Figure VI (photo section) is Cygnus A, a faint irregular galaxy located in the constellation of Cygnus. Aside from its shape there is little in the appearance of this galaxy to excite attention. Nevertheless, it is one of the most powerful emitters of radio signals in the sky.

These radio signals do not come from the visible galaxy itself. They come from very far away from it—from two vast regions of space lying on either side. The radio-emitting regions are symmetrically placed about the galaxy, and each is very much larger than it. Also shown in Figure VI is a map of the observed radio emission.

Perhaps the strangest thing about these vast areas is that to the eye they appear quite empty. Detailed searches of photographs (such as Figure VI) for visible objects have been in vain. Apparently, nothing an optical telescope can reveal serves to distinguish these expanses from any other region of intergalactic emptiness. It takes a radio telescope to show that they are full.

The two regions are full of electrons and a magnetic field—the magnetic field lines a giant tangled skein, twisting this way and that, extending over millions of light years; the electrons of enormous energy and velocity, shooting through the skein, trapped by it, arcing in complicated paths along the magnetic lines of force. Just as in the pulsars and the Crab Nebula, it is this combination of rapidly moving electrons and a magnetic field that produces the radio emission. So vast are these clouds, so energetic are the electrons, that the radio emission is stronger than the visible emission of light from all the hundreds of billions of stars in Cygnus A added together.

It is difficult to escape the impression that these electron clouds were shot out of Cygnus A. They are both the same distance from the galaxy. Two weak bridges of radio emission are clearly visible in Figure VI extending from the clouds toward the galaxy, and it may well be that larger radio telescopes will show these bridges extending all the way down to the visible galaxy. As for this galaxy itself, it is irregular in shape. One wonders whether some cataclysmic event may not have taken place within it millions of years in the past.

Many galaxies are irregular, and most show no traces of an explosion. But Cygnus A has another peculiarity as well. Recent observations with ultrahigh resolution have revealed additional radio emission from a single, intense knot lying at the galaxy's nucleus. And most significantly of all, this knot is located exactly midway between the electron clouds.

The state of affairs is surprisingly similar to what has been found in M87. For all their differences Cygnus A and M87 are alike in two essential respects: each contains large, diffuse regions of radio emission; and each contains a knot, a sharp pointlike center of activity at its core. By analogy, it may be that Cygnus A too contains a giant black hole.

Even without the analogy many scientists have been led to this supposition. The electron clouds must have come from somewhere, and if it was an explosion that produced them it must have been inconceivably huge. Beside it a thermonuclear bomb would be utterly insignificant in comparison, a popgun. Even the supernova fades to insignificance beside this ancient explosion. So powerful must it have been that one is hard pressed to explain it. It is almost too big for comprehension. By necessity, many workers in the field have come to the conclusion that only some explanation involving a giant black hole at the nucleus of Cygnus A has a chance of accounting for its properties.

Figure VII (photo section) turns to still another scene of violent activity. The arrow in this figure appears to point to a star. But it is not a star—it is the quasar 3C48. That tiny pointlike spark lies 100 million times farther away from us than the star beside it on the photograph. The quasars are the most distant objects known, and they are so extraordinarily far away that their light has taken fully billions of years to reach us. The faint gleams of light from the most distant were emitted in ages so prehistoric, so long gone, that life itself had not yet come into being on our own world. Figure VII is a snapshot of the universe when it was young.

The only reason such distant objects can be detected is that they are very bright. The quasar, which looks like a star, is one thousand billion times brighter than a star: one hundred times brighter than an entire galaxy. It is the most powerful object in the universe. Nothing else known to science approaches the quasar in the intensity of its emissions.

Nevertheless, they are very small in astronomical terms. A typical quasar might be a millionth the size of a galaxy. This enormous degree of compression is one of their most significant features, and it immediately brings to mind the black hole. Black holes naturally produce such tight, compact blazes of emission as they absorb matter.

There is a further significant feature of quasars that fits in with this interpretation as well. It is their variability. In contrast to the radio galaxies, quasars do not shine steadily. They suffer flares, explosions: monstrous, catastrophic bursts in which the already huge emission rate can double within a month. For all the world the quasars appear to be one long chain of cataclysms. Such behavior is roughly what would be expected if they did consist of giant black holes into which stars fell at random. Each encounter would release the sort of explosive burst that is observed.

From the farthest reaches of the universe we return to our own backyard—to the Milky Way Galaxy in which we live. The sky above shows no evidence of giant explosions. Terrible blasts do not rain down daily from the sky. Nevertheless, recent discoveries have shown that the nucleus of our galaxy is a scene of intense activity. The winding spiral arms extend all the way down into this region—but they are expanding there, rushing outward at great velocities. A ring of interstellar molecules has been found, a doughnut-shaped cloud of emission that appears to have been pushed out from the center. It is expanding at a quarter of a million miles per hour. Outside this ring is an ionized cloud that both rotates and expands. A tiny knot of radio, X-ray, and infrared emission has been detected there. All in all, the output from the galactic center is equivalent in power to that of a hundred million suns.

Are the various structures at the heart of the galaxy expanding so rapidly because they were blown outward? Was there a giant explosion within the Milky Way far back in the past? The first thing to be said is that our galaxy cannot legitimately be classified with the radio galaxies, for it possesses no traces of those giant electron clouds. Nor is the emission from its nucleus remotely comparable in strength to theirs. Furthermore there is no evidence on the Earth that such an explosion ever took place. It would have occurred about a million years ago. The fossil record is very complete from this period, and it shows nothing unusual. No widespread extinctions occurred at that time. (The dinosaurs died out not one but 70 million years ago.) Fossilized ocean plankton is perhaps the most sensitive tracer geologists have found of ancient mutation rates, and it does not show the sudden jump one would expect from cosmic radioactivity produced by such an explosion. Nor was there any major change in weather. The period around one million years ago coincided with the peak of the ice age, and witnessed the farthest incursion of glaciers into Europe and North America. But the ice age is still going on—we are simply in the midst of a brief respite at present. Furthermore, the glaciers expanded southward not once but four separate times, and it is hard to see how a single explosion could have caused this.

All in all, the evidence is inconclusive. If there was an explosion in the nucleus of our galaxy, it must have been a relatively small one which left no permanent traces on either the Earth or the galaxy as a whole. Nevertheless the activity in the core remains to be explained. There is no disputing the fact that the nucleus of the galaxy is the site of intense emissions and various expanding structures. Something is producing this activity, and that something may possibly be a giant black hole. Dark, invisible, millions of times the mass of a star, the hole may lie brooding at the galaxy's heart.

Radio galaxies, quasars, our own Milky Way . . . everywhere we look we find possible evidence of black holes. And if so, we are led to a very interesting speculation indeed.

For what, after all, is a galaxy? It is a swarm of stars. For decades astronomers have believed that it is nothing but a swarm of stars. But perhaps this is mistaken. Perhaps a galaxy is *a swarm of stars surrounding a black hole*. Perhaps every galaxy contains at its heart a giant hole.

It may even be that the black hole is the important thing. It may be that the blackness of the hole has masked until now its presence, and the brightness of the stars led us to overemphasize their significance. Possibly the true meaning of all the billions upon billions of stars comprising a galaxy is that they are merely markers, signposts signaling the presence of a hole.

In this view radio galaxies and quasars are not so very different from our own galaxy at all. All contain black holes of comparable size, the only difference being the degree of activity of the hole. Within ordinary galaxies such as our own, they are more or less quiescent—presumably because there is relatively little matter in their vicinity to be drawn in. Within radio galaxies and quasars, on the other hand, the hole is in close contact with large amounts of matter and produces violent effects. It is even possible that every galaxy regularly passes through such violent stages, and that such activity is part and parcel of their normal evolution. If so, then life in the universe would be much more rare than we had thought, for each time a galaxy underwent such a cataclysm every living creature it contained would be killed. Galaxies would regularly purge themselves of life, as forests are periodically cleared by forest fires.

If all this is so, astronomers would have at their disposal the answer to a vexing question. It is the question of the origin of galaxies. Galaxies, in fact, have always been mysterious things, and no one has succeeded in explaining where they came from. The traditional view is that the universe originated in the Big Bang, and that in this explosion a purely uniform, homogeneous cloud was created. Out of this expanding primeval stuff condensations are thought to have formed, enormous clumps billions of times the mass of a star. Initially they could hardly be said to have a separate existence, and were merely clouds of gas slightly denser than the average. But as time passed, gravity is thought to have pulled them together and made them into more coherent units. Ultimately these condensations themselves broke up into subcondensations—billions of them, each the mass of a star—and ultimately these microscopic blobs contracted into stars.

Such is the traditional view of the origin of galaxies. and it is a grand view indeed. The trouble is that no one has ever succeeded in making it work. The difficulty is that the universe expands from the Big Bang too fast. It expands so quickly that it drags apart the condensations as they form. The expansion of the universe acts against the compressing effects of gravitation; and rather than contracting into separate, identifiable units, the initial protogalaxies are pulled apart. They never get a chance to form.

But if galaxies really contain giant black holes, this problem can be solved. In this case it is the intense gravitational pull of the hole that draws together the protocloud—and this gravity is so strong that it succeeds in counteracting the effects of the expansion of the universe. Rushing outward away from the Big Bang, we imagine a swarm of black holes mixed together with a uniform cloud of gas. Around each hole the gas clusters to form galaxies. The galaxies form because of the holes.

Perhaps so. But many do not agree. Many astronomers do not believe that there is a black hole in the nucleus of our galaxy. Nor do they believe that black holes exist in radio galaxies or in the quasars.

These astronomers point out that in each case the evidence for the existence of a hole is tenuous. The best case we have is the galaxy M87, and it is going to take the space telescope to finally decide that question. In the Milky Way Galaxy the situation is ambiguous and is likely to remain so for some time. Radio galaxies like Cygnus A and quasars like 3C48 are so far away that there is no hope of ever directly observing the hole. For such objects we are only left with circumstantial evidence—signs of explosions, and unusual activity in the core. Who can be sure that it takes a black hole to produce such things?

Many alternative explanations have been proposed. One group of explanations hypothesizes that occasionally in the universe there may be found small, ultradense clusters of stars. So dense could these clusters be that from time to time stars would actually collide within them. The collisions would be so cataclysmic that both stars would be destroyed—blown to bits—and the sort of flare observed in the quasars would result. Other explanations propose the existence of giant rotating clouds millions of times the mass of the Sun: super-pulsars.

We are speaking here of places so far away, of objects so vast and unfamiliar and strange, that our ignorance is profound. It is *too* profound, and we are lost in the realm of speculation. For every theory proposed there is another theory that contradicts it. For every discovery there are a variety of possible explanations. I myself agree with the skeptics: in my view no one has succeeded in presenting incontrovertible evidence for the existence of a single giant black hole.

As for smaller holes, holes the mass of one star, that is another matter. It is discussed in Chapter 14.

The black hole's force of attraction comes from some *thing*—from an object. It does not come from the hole itself; the hole is simply a result of this attraction, an area of darkness arising from its effect on light. Transfer attention now away from the hole, and onto the object that makes it. How shall we describe it? What sort of object could it be?

With this simple question, so innocent in its appearance, we are plunged into a mystery. We have reached now one of the most profound of all the enigmas of modern science. More than that—physics reaches a crisis. It is no exaggeration to say that in the answer to this question lies the solution to one of the most ancient of all riddles: the riddle of the ultimate nature of matter. For according to the Schwarzschild solution, the object responsible for the black hole in a very real sense cannot exist. It is crushed into nothingness.

It is crushed into nothingness by gravity—by the very same infinite, irresistible force of attraction that everything feels that enters the black hole. Things within a hole fall helplessly. So does the star that makes the hole. It falls inward upon itself.

Every star in the sky exists in a state of compression produced by gravity. The Sun experiences this force right now. Each particle within it exerts a gravitational attraction on each other particle: although the attraction of one atom for another is small, there are so many of them that the total force is very large. And the net result of all these myriad particles attracting and being attracted by each other is a powerful, inward-directed force of compression.

In ordinary situations this force is resisted by another—by the pressure at the center of the star. So hot is the star's core, so great is the pressure there, that it is sufficient to oppose gravity. The star hovers in the balance of forces: gravity compressing it inward, pressure expanding it outward.

But gravitation within a black hole is irresistible. The star within the black hole—the star which makes the black hole—experiences an infinite compressional force. No countervailing pressure in its core can resist this: no matter how hot the star, no matter how enormous the pressure, gravity wins. The star falls like a skyscraper whose supporting beams have given way—except it falls not *downward* but *inward*. It implodes, and it does so at the speed of light.

In this collapse matter is rushed through increasing degrees of compaction. Even when a star is larger than its Schwarzschild radius, it is denser than an atom. The atoms out of which it is made overlap. As a result they dissolve into their constituent parts. The star is composed not of atoms but of electrons and nuclei. It collapses inward. Within a fraction of a second it has reached the dimensions of a pulsar—it has become a pulsar. The nuclei are forced together. Their edges touch. An enormous pressure develops as they resist the compression and seek to maintain their structure. But although this pressure is large, it is not infinite. Gravity overwhelms, and the nuclei are rammed together. Now it is they that dissolve, into protons and neutrons. The collapse proceeds. Within a fraction of a second the star has shrunk until it is smaller than a pulsar. It has become an object for which we have no name—an object the likes of which no one has ever seen before. The very protons and neutrons are compressed together. The elementary particles dissolve.

Such a situation also holds in the core of a neutron star and, like there, the material may shred into quarks. But there is a difference. The difference is that this star is still collapsing. It has reached in its collapse a degree of compression as great as anywhere else in the universe—but it does not stop there. It keeps right on contracting, rushing together at the speed of light, and in a fraction of a second the quarks are forced together.

Now the collapse has gone beyond the edge of knowledge. It has crossed the frontier of modern physics into wholly unknown territory. Elementary particle physicists do not have the slightest idea what quarks are made of. Conceivably, however, they are made out of *something*; and whatever this something is, the quarks dissolve into it. Now the star is made out of elementary particles of whose very existence we are ignorant. And the collapse continues.

Where does this all end? How long can matter be crushed together in this way, passing again and again into new stages of composition? What is the final endpoint of the collapse?

There is no final endpoint, and on this general relativity is absolutely clear. It is here that physics reaches its crisis. It is one thing for the collapse to go a step or two beyond our present understanding—to the next few levels of the structure of elementary particles. It is another matter altogether to go all the way to an infinite compression. But this is actually where the collapse leads. It leads to a final state in which the star is not a mile across, not an inch, not even the dimensions of an electron, but *of zero size*. It reaches this state necessarily and inevitably, and it does so in a fraction of a second.

The collapsing star is made of matter but in the collapse matter's very nature is called into question. For what is substance? What is "stuff"? It is not easy to give an answer to this question. Everyone knows what matter is, but when we try to express ourselves the certainty can slip away. At least some of the time, stuff is solidity. If I bump into a wall in the dark I have encountered matter. It resists me—it hurts. If I drop a brick on my toe I feel another property of matter: its weight. As I maneuver around a trunk in the crowded attic I experience still a third property: matter occupies space. And lastly, matter is something that can be seen.

But not all these are universal properties of matter. Air is matter but air is not solid: no one ever banged against a piece of air. As for weight, far out in space it is absent. Things are weightless when distant from the Earth. And plenty of things are invisible—atoms, for instance.

Can we refine our thinking? Are there any properties of matter that are truly fundamental and shared by all objects?

Although it is true that air never bumps into things, it is also true that air can push things around. Fight against a powerful wind and you will convince yourself of that. A tornado—pure, insubstantial air—can lift buildings bodily upward. Where does this power come from? In the ultimate analysis it comes from the *inertia* of matter. Moving air—wind—carries inertia, just as a moving car does; and anything that resists this motion tends to be brushed aside. It makes no difference whether the object in motion is a solid or a liquid or a gas: so long as it is moving, it requires a force to slow it down, and it can exert forces on other things by virtue of this motion.

Physicists measure the quantity of inertia possessed by a body through its *mass*. Mass is not weight. An astronaut far out in space may be weightless, but he has exactly the same mass that he had back on Earth. On the Moon, where he weighs one-sixth his normal weight, his mass is also unchanged. It never changes: mass is one of the universal properties of matter.

Another universal property is extension. Things occupy space. *All* things occupy space. A suitcase cannot carry an infinite number of objects. This is true in daily experience; it is also true in physics. No matter how tiny the thing considered, it still has some definite size. Even the elementary particles are not infinitely small. The electron, the smallest particle known, is about $1/10,000,000,000,000$ of an inch in diameter, and although this may seem like zero, it is not. There is a very definite limit to the number of electrons one could pack into a suitcase.

Mass and size: these are the fundamental properties of matter. Indeed, many physicists would *define* matter to be that which has inertia and extension. But inside the black hole one of these two properties is extinguished.

What is matter like that has no size? What would it look like if I could see some? In my hand I hold a spoon. It is made of metal, plastic, wood—whatever: it is made of *stuff*. I put it in a vise and crush it. Now the spoon is out of shape: it no longer is a spoon. But the matter out of which it is made is still there. I put it in a blast furnace and melt it down. I vaporize it and blow away the gas. But in all these transformations the matter, the pure, primordial substance out of which the spoon is made—this remains. But what if I crush the spoon till it has no size at all? Does the matter even exist anymore?

To these questions no one has an answer. None of us can conceive of such a thing. Neither can physics help. Nothing that science has learned gives the slightest clue to what happens when gravitational collapse has progressed to its ultimate end.

I will tell you what something looks like that has no size. It looks like a black hole.

Physicists have a word for the thing that lies at the center of the black hole. They call it a singularity. A singularity is a point where your theory goes haywire. The mathematical function $1/x$ has a singularity when x is o. One divided by o is not just infinite—it is singular. It is an illegal operation in mathematics to divide by o: not allowed. But this is just what the Schwarzschild solution involves. Inescapably, necessarily, anything that grows too small suddenly implodes to a singularity; anything that approaches too close is pulled in too. And at this singularity everything breaks down. Physics comes to an end at the center of the black hole.

No one knows what to do with the singularity. It is not just a theory of the internal constitution of quarks that we need. We need to know whether matter has to have a size—whether extension is as fundamental to its nature as mass. But even more than this will ultimately be necessary, for general relativity itself breaks down at the singularity. The very theory which predicts its existence founders upon it. General relativity self-destructs.

What we require is some major modification of general relativity. Einstein's work stands in need of revision. Most physicists believe that a unification of relativity with quantum mechanics will point the way to a resolution of these mysteries. But this unification eludes us. No one has succeeded in bringing it about. After decades of effort, relativity and quantum theory, the two major creations of twentieth-century physics, remain separate, unrelated edifices. And until they are joined the singularity will remain.

11

Spacetime Geometry

In the two previous chapters a light bulb was lowered into a black hole. Now lower something different: lower a kitchen! An entire room dangles on the end of the rope. In this room a man is standing. I intend to study him through a pair of binoculars as he moves about.

The scene that greets my eyes is one of abnormal dimness. Also, its color is wrong. The room is illuminated by ordinary electric lights but they do not seem to be working very well. They are no longer a bright and cheerful yellow but dull red in hue. Through this oppressive reddish gloom the man is walking toward the sink. In his hand he holds a saucepan.

He is walking very slowly. Every step takes a great amount of time. He is not striding across the room—he is ambling, creeping. Eventually he reaches the sink. With a languid gesture he reaches forward to turn on the water. The stream from the tap moves with lugubrious slowness. It does not shoot out—it crawls. The water flows like molasses.

Everything I am seeing looks like a movie shown in slow motion. The saucepan takes too long to fill. The man takes too long to carry it over to the stove. It seems to take forever to bring the water to a boil.

He is cooking an egg! Into the water the egg is lowered, carefully, deliberately. The egg timer is set. There must be something wrong with it for it runs too slow. After a full five minutes have passed it has advanced by only one. Unconcerned, the man sits reading a newspaper. Clearly he is a slow reader. He lingers over every page. Ten minutes pass, fifteen. He never shifts his weight.

Finally the timer pings. The man pulls the egg out of the water and cracks it open. Onto his plate falls a perfectly cooked three-minute egg.

It is tempting to suppose that these strange things are caused by the terrible force of gravity near the hole. After all, that man was fighting an enormous gravitational pull when he walked across the room. It must have slowed him down. And the egg timer is composed of some physical mechanism—levers, springs. Might it have been warped, slowed by gravity?

A little thought shows that this is not going to do. The egg timer's abnormal behavior might be accounted for in this way but certainly not that of the egg. A pure and simple force, no matter how intense, has no way of retarding the myriad chemical reactions that go into its cooking. And how can gravity make water flow so slowly? How can it change the color of light?

It is not events that are slowed. It is time itself. The man was not creeping across the room at all. He was moving at a perfectly ordinary pace. Water was flowing from the tap in its usual hearty gush, and the egg cooked at the normal rate. Every one of these processes took place *in time*—and it is to this passage of time rather than to the processes themselves that we must look. Close to a black hole time passes more slowly. It passes more slowly relative to its rate far from the hole. The stronger the gravitational field the more the passage of time is retarded.

This also explains the reddish cast of light that had filled the scene. The physiological sensation of color is triggered by a light wave of a certain frequency: the lower the frequency, the redder the light is perceived to be. But frequency in turn is a certain rate of vibrations per second. If the rate is slowed we perceive a redder color.

So long as I remain fixed in one place and confine attention to my immediate surroundings, nothing out of the ordinary will result. I will believe that time is passing at its usual rate. This remains true no matter how close to the hole I suspend myself. I can verify this by shinnying down the rope to join that man in the kitchen: once I arrive there, everything strange that I had seen from a distance goes away. Standing there in the room I notice nothing unusual about the passage of events. Things weigh a great deal, to be sure, and the paths of light rays are bent, but time is certainly not affected. Clocks tick along at the rate of one second per second, and a soft-boiled egg takes just three minutes to be done.

It takes three of *my* minutes. But these are different things than the minutes proper to another place. Once I transfer my attention away from my immediate surroundings and study things at a great distance, time becomes desynchronized. If I gaze downward and study events occurring even closer to the black hole, I see them occurring with the same exaggerated slowness that I had noticed before. If I gaze *up*, away from the hole, I see just the opposite. I see events proceeding far too rapidly. People high above my head dash about madly. They seem to be incessantly in a rush. One man chops down a tree in an instant. It takes a fraction of a second to fall. Another hurriedly thrusts an egg into a pan of boiling water and jerks it out again right away. He cracks open a three-minute egg and devours it in a trice. Everything going on up there is suffused by an abnormally brilliant glare, and it is not soft yellow in color but a harsh electric blue.

The closer I drop toward the horizon, the faster things high above seem to go. I can lower myself until centuries pass in an instant. I can sit in my armchair and gaze upward at the future history of the world. Empires come and go. The United States of America is reduced to an unimportant bystander in the ebb and flow of world events. Continents drift apart so rapidly I can see them moving with the naked eye. The Himalayas are worn down to rolling hills and a new mountain chain is thrust up in Kansas.

There is no limit to this process. By hovering a fraction of a millimeter above the black hole's horizon I can see the galaxy turning ponderously. I can watch the very expansion of the universe itself. And as I lower myself *through* the horizon, the entire future history of the cosmos flicks by overhead. It is completed just as I cross into the hole—just as I lose control and plunge downward to my death upon the singularity.

A person situated high above and given the task of studying me would see me slowing to a virtual halt in my passage toward the hole. This person would die of old age before I had drawn a breath. His son might take on the task of watching me. A trust might be established to ensure that future generations would maintain the watch indefinitely. Each generation would see me ever closer to the hole, ever redder, ever dimmer, always moving more slowly. From their perspective I would *never* cross the horizon. The Sun would burn out and die, the human race would come to an end . . . only one survivor would remain: entombed, frozen, trapped in time beside the hole.

I have been speaking far too glibly about black holes so far. I have not been sufficiently careful. A black hole is formed when a star collapses inward to cross its Schwarzschild radius. But is such a thing even possible? How long does it take?

If there is anything relativity teaches, it is that we must be careful when speaking about such things. We must decide where we are standing when we ask the question. So make a choice. Begin by standing *on the surface of the collapsing star.* I will ride with the star down into oblivion.

In this frame of reference the collapse is very rapid. It takes about two hours for the star to fall inward from its present great size to the point at which it is just outside its Schwarzschild radius. As far as I am concerned, it takes a mere fraction of a second thereafter to cross this critical dividing line. Now the black hole has formed and I am inside it. So is the star beneath me. We both plunge into the singularity in an instant.

Now change the frame of reference. Stand upon *the Earth* and watch the collapsing star. From this vantage point things are very different; the collapse goes on forever.

Initially there is no difference between what one sees and what I on the star experienced, for gravity on its surface is too weak at first to slow the passage of time. During the first two hours of the collapse this remains true. But as the star approaches its Schwarzschild surface the queer desynchronization of time begins to set in. The collapse appears to slow. Soon the star appears to be hovering, floating just above its Schwarzschild surface. Dull red in color, vague, dim, endlessly creeping inward . . . it has become a ghost.

But it has not become a black hole. It never will.

Soviet scientists have coined a term for such an indefinitely collapsing configuration. They call it a *frozen star.* And the striking thing about frozen stars is that *they exist. Black holes, on the other hand, do not exist.* There is not time enough to form them.

In another sense, however, black holes are very real indeed. *The way to form one is to jump into a frozen star.* As I fall onto an endlessly collapsing star the flow of my time is stretched to match its own. It brightens and grows more white, and it resumes its rapid infall. I fall upon the star, it falls within its Schwarzschild radius, and we both enter a black hole.

So long as we remain upon the Earth and search for such objects with telescopes, we will not be troubled by these ambiguities. What we are looking for is frozen stars. Although a frozen star is different from a black hole in principle, there is no difference between the two in practice. So strongly is the light from a frozen star sucked down by gravitation that it is quite impossible to detect. It would be literally invisible to the naked eye at a distance of 100 yards. Chapter 7's description of an encounter with a "black hole" paints a perfectly accurate picture of what a frozen star would look like. For this reason relativists speak of black holes and frozen stars almost interchangeably. Everything written here of black holes applies equally well to frozen stars. Only if by some means we could succeed in capturing that one photon per century emitted by the indefinitely collapsing star would we be able to distinguish it from a hole.

Not only is the passage of time altered by gravity. So too is the nature of space. And the way in which relativity treats these things is in the unified language of space-time.

Begin with an illustration: an automobile journey due north that lasts just an hour. On a map this journey would be illustrated as a straight line running from south to north. But there is a second sort of illustration that could be drawn—one that includes a description of the time required for the trip. Graph the distance traveled against elapsed time as in Figure 37.

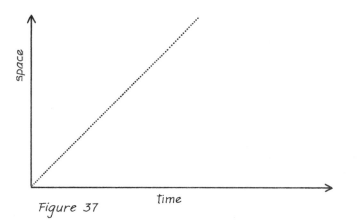

Figure 37

This is the simplest version of a *space-time diagram*. Clearly there is nothing very strange about it. Equally clearly, it contains more information than the first method of representation. Suppose for instance that the driver had stopped for a while at some point during the trip. The route as illustrated on a map has no way of representing this stop, but in the space-time diagram it is illustrated as a kink in the line (Figure 38).

The space-time diagram adds another dimension, that of time, to a description of a process. A one-dimensional automobile trip—a trip due north—requires two dimensions for its space-time representa-

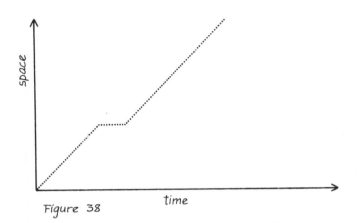

Figure 38

tion. A two-dimensional trip needs three. A path in full three-dimensional space would require four dimensions to be represented in this way. Of course we cannot draw such a picture. But nothing prevents us from thinking about it.

What is the significance of the space-time diagram? So far it has none whatever, it is simply a way of illustrating things. Einstein was certainly aware of it early in his career but he had given it no particular attention. But after developing special relativity he came to change his mind on this. Far from being a mere convenient fiction, a method of illustration, space-time became more and more significant to him. Eventually he came to the view that it was *real*, as real as tables and chairs. Einstein decided that the true arena in which physical events ran their course was not space and time, but space-time. It was a four-dimensional world we lived in and the theory of general relativity that he went on to develop is exclusively couched in these terms. The proper language in which to discuss black holes is that of space-time.

High on a mountain a stone is dislodged from its perch. It falls. Initially it falls in a straight line and at an ever-increasing rate, but soon it strikes against the cliff. The collision deflects the stone's path outward and briefly slows its rate of fall. Again and again it strikes, the velocity and direction changing with every blow.

A translation of this description of a falling stone into space-time language requires a diagram of four dimensions—up and down, forward and back, right and left, and past and future. In this diagram the fall is represented by a four-dimensional tube whose "thickness" is the three-dimensional volume of the stone. The tube twists and turns as it snakes through the diagram. Its primary course lies from "up" to "down" along one space axis, and from "begin" to "end" along the time axis, but from point to point the tube makes sharp bends in the other two space directions as well, each kink representing a collision of the stone.

The space-time description of the full physical world would be a diagram of almost inconceivable complexity. It is filled with tubes of varying width—thick ones representing large objects like cars, thin ones representing smaller things such as individual molecules. They tangle about each other in complicated ways. A bundle lying along the time axis for a while represents cars waiting at a red light: at the point in the diagram at which the light turns green they angle off and diverge somewhat as the cars separate. A traffic accident is represented as the intersection of two tubes; birth as the branching of a smaller from a larger one. And all these myriad tubes are wrapped like vines about a larger one representing the motion of the Earth itself. It is a curiously static picture of things. Nothing ever *happens* in the four-dimensional world of space-time. All the past and all the future are displayed at once. The universe simply is.

Among all the laws of physics the principle of inertia is surely one of the simplest. It says that objects do not like to change their state of motion. More precisely, it says that a force is required to bring about this change. If no forces act on an object, then its state of motion continues forever. If initially at rest it remains at rest: if initially in motion it continues moving in the same direction and with the same speed.

In this case the space-time diagram of the motion will be quite simple. The tube representing the object's path will be a straight line. This can even be expressed as a law of nature if we like. The principle of inertia can be reformulated into four-dimensional language to say that *the space-time diagram of an object on which no forces act is a straight line.*

So much for motion in the absence of forces. Now consider an object under the action of a gravitational field. To make things specific let the gravity come from the Sun, and the object in motion be the Earth itself. The Earth orbits in a circle.

Now draw the space-time diagram of the motion. As the motion along the time axis advances by one year, the motion in the space axes swings about in a circle. So the diagram is a corkscrew as in Figure 39.

This is certainly not a straight line—or at least it does not look like one. But here we are mistaken. Strange as it may sound, that

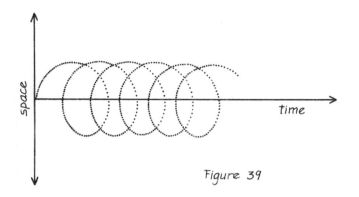

Figure 39

curving spiral is in fact perfectly straight. According to Einstein, gravitation is not a force at all—it is a distortion of the nature of spacetime. In this view there are no forces whatever acting on the Earth in its orbit about the Sun, and it moves just as the law of inertia says it must: in a straight line. But the line is in a new geometry.

How can there be such a thing? How can geometry be distorted? We are accustomed to thinking of geometrical statements as fixed and inflexible. Also, we are accustomed to thinking of them as *true*. Who could doubt that a straight line can be continued forever, or that parallel lines never meet? So firmly ingrained are these things in our minds that the very thought they might be wrong seems like madness.

Nevertheless, it is worth asking just why we are so sure. What gives us this massive certainty that Euclid was right? Has anyone ever checked? Has there ever been an experiment to see if the real world actually conforms to Euclidean geometry?

In crude terms it happens every day. If parallel lines were to meet we would be in serious trouble every time we drove down the road. What would happen when we passed a truck heading parallel to us in the opposite direction? Nor does it seem on first glance that the hypotenuse of a triangle is longer than the sum of its other two sides—if it did the long way around would really be the short. But these are not really such precise statements. After all, perhaps the cars on the other side of the road are not heading exactly parallel to us, and perhaps the triangle does not exactly conform to the Pythagorean theorem. Perhaps it might make sense to check more carefully.

One of Euclid's most famous propositions is that the circumference of a circle is π times its diameter. To test this a real, physical circle must be built and its dimensions measured. The following method will do. A circle is defined as the set of all points equidistant from a given point. For the given point—the center of the circle—drive a stake into the ground. For the radius of the circle take a rope. One end of this rope is tied to the stake, the other around my waist. I walk in a circle, straining against the rope (Figure 40).

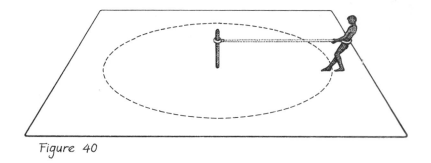

Figure 40

Now I measure the distance I have walked and divide by twice the length of the rope. If I pay out 10 feet of rope and carry out the operation, I find the circumference to be just over 60 feet. The measured value of π works out to a bit more than 3. It is close but not right on. I will have to measure things more exactly. Nothing prevents me from doing so—I don't use something crude like a yardstick, but rather something finer. Doing this I find the measured value of π to be 3.14159, and the more accurately I measure things the more decimal places I will be able to add on.

The result is in agreement with the predictions of pure mathematics. Euclid has won an important victory. But wait—I am not yet done. Euclid claims that *all* circles have this same ratio of circumference to diameter, no matter how small or big they may be. Test this statement. Draw a bigger circle.

Pay out the rope till it is 100 miles long. I walk in an enormous circle. The distance I have trudged before returning to my starting point adds up to 628.252 miles. But now I am in for a surprise. Now the measured value of π is not 3.14159 at all. It is 3.14126.

Something has gone wrong. Have I made an error of measurement? No matter how carefully I repeat my observations I get the same result. Apparently there is nothing for it but to go on. Pay out the rope till it is 1,000 miles long. My circle is now 2,000 miles in diameter, its circumference more than six. The measured value of π that I obtain is 3.10876.

The bigger the circle grows the smaller π becomes. When the radius of the circle is 6,261 miles, π is just equal to 2. Beyond this point it is not just the ratio of circumference to diameter that shrinks: the circumference itself begins to drop. The bigger the circle the smaller its circumference. Smaller and smaller grows the path in which I walk. When the rope is 12,500 miles long the circle I pace out is little more than 100 miles around; and finally, when the radius is 12,522 miles, the circumference has shrunk to zero. The value of π, as determined from measurements performed on this huge circle, is not 3.14159, it is 0.00000.

How are we to deal with this strange state of affairs? How can we understand a shrinking value of π? We do so by recognizing that the circles have been drawn *on the surface of the Earth*. And the Earth is round.

Draw a picture of the operation I have been performing. For a small circle, Figure 40 will do, and there is nothing strange here. But for a bigger circle another element if the situation comes into play— an element not shown in that small scale picture. It is the curving bulk of the Earth itself. Large circles were constructed as shown in Figure 41.

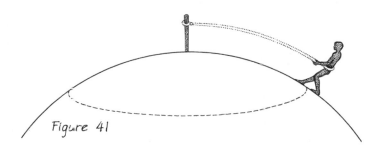

Figure 41

And this is not the sort of situation Euclid had in mind.

Toward the end of the experiment the rope defining the circle's radius was stretching far across the curve of the Earth. Ultimately its length exceeded one-quarter of the way around the world—and from then on strange effects were bound to come into play. In this situation the circle actually shrinks as the rope is paid out. And finally, when its length is just half the Earth's circumference, the circle has shrunk to oblivion (Figure 42).

The temptation is strong to conclude that Euclid is correct after

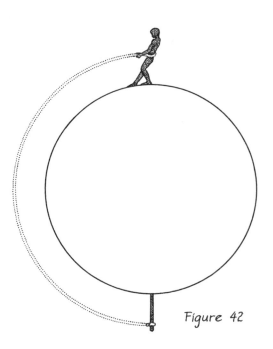

Figure 42

all. But examine this more carefully. Why the sense of relief? Who is to say that the arcing path defined by the rope is not the true radius of the circle?

Very likely the following answer will come to the reader's mind. "I recognize full well that in the imaginary experiment you were straining against the rope, and that in ordinary conditions this is enough to guarantee that it be straight. Unfortunately in this case it was not sufficient. Since the radius of a circle must be a straight line the experiment contained an error."

I am immediately inclined to accept your answer, but a nagging question remains. What do you mean by straight? And how do you find out whether a line is straight?

These are not such very easy questions to deal with. What, after all, do we mean by a straight line? Let us turn to Euclid for help. Does he say what he means when he speaks of one? Here is what he has to say:

> A *straight line is one which lies evenly between two of its points.*

Translated into modern language, this says that a straight line has no kinks in it. It does not bend. But modern mathematics has found this a difficult definition to accept. It is not so very easy to say rigorously what you mean by a bend. Because of this many mathematicians nowadays are inclined to adopt a different definition of straightness.

A straight line is the shortest distance between two points.

In some situations these two statements are equivalent. If we are concerned with lines drawn on a flat piece of paper, they both add up to the same thing. But if we are drawing lines on the surface of a sphere, they do *not* add up to the same thing; and indeed, as far as the surface of the Earth is concerned there is no such thing as a line satisfying Euclid's definition of straightness. On a curved surface all lines bend.

On the other hand, there certainly is a path that can be drawn satisfying the more modern definition. This path is a great circle: the intersection of the surface of the Earth with a plane passing through its center. The equator is a great circle, and so are the meridians of longitude. In the previous experiment the rope followed one of these meridians—that passing between the pole and me. It defined the straight line connecting these two points—I forced it to by pulling it tight.

Therefore by the modern definition of straightness the rope *did* represent the true radius of the circle. The experiment was legitimate after all: on the surface of the Earth geometry actually is non-Euclidean.

There is a term for all this. The term is "curved space," and it implies that a non-Euclidean geometry was created by the simple act of redefining the problem. Euclid wanted to talk about figures drawn on a plane. If we decide to do so with him all well and good: we will agree with his results. But we might also want to bend that plane into a curve, and if we then decide to draw figures on it there should be no surprise if we get something new. The curved-space interpretation of non-Euclidean geometries holds that they differ from Euclid's simply because they are speaking about *something else*.

But what if we do not want to change the rules? What if we go along with Euclid and agree to draw lines on a flat plane? Is it *still* possible that his geometry is in error? This is a far more profound question than the curved-space interpretation would allow. It is asking if the geometry of "flat" space—of real space—might be non-Euclidean.

The first thing to be said is that no departures from Euclidean predictions have ever been observed. Actual experiments similar in spirit to that described above have been done, but so far have been inconclusive. But this does not resolve the issue since these departures may be very small. On the theoretical side it is clear that any attack on Euclid's work will have to find a weak point in his argument. This weak point is not to be discovered in his proofs themselves. They are models of correctness. Rather it resides in the very foundations of his theory—in the definitions, axioms, and postulates he chose to adopt. And the striking thing about these is that Euclid made no attempt whatever to justify them. To him they required no justification: they were self-evident. It is on these unproven assumptions that the entire structure of his geometry rests. They are its only weak point.

Among these assumptions one from the very outset had seemed suspect. It is called the *parallel postulate*, and in modern language it states that if we have a line and a point not on this line, then one and only one new line can be drawn through the point parallel to the given line. Draw the picture. We are given Figure 43

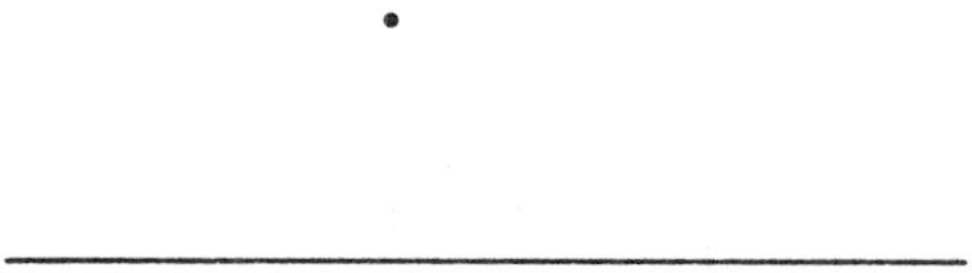

·

_____

Figure 43

and we are asked to draw a parallel through the point. Can we do it?
Figure 44 gives it a try.

— — — — — — — ◆ — — — — — — — —

_____

Figure 44

It certainly seems as if we have succeeded. But have we? How to make sure that the line we have drawn is parallel to the old?

The test is to prove that the two lines never meet. Now "never" is a strong word. They certainly do not meet on the figure as it is drawn. On the other hand, neither do those of Figure 45.

_____

Figure 45

and they certainly are not parallel, as can be seen by extending them until they cross. Similarly, in order to show that our supposedly parallel lines really are parallel, they must be extended—*forever.*

Clearly this is not an operation that can actually be carried out. No matter how far the lines are extended they can always be extended some more. Paper can be glued together to make a sheet 100 miles long and their extensions drawn on it. If they intersect we have shown they are not parallel—but if they do not intersect, we have not shown that they are. Perhaps they will cross in the next 100 miles—or 100,000,000.

It was this "endless" quality of the parallel postulate that troubled mathematicians after Euclid. It did not seem so clear to them that the postulate was even true. Who could say what happens infinitely far away? Accordingly mathematicians sought to prove the postulate. They sought to deduce it from the other definitions, axioms, and postulates of Euclidean geometry. This effort went on for centuries and it was utterly unsuccessful. Many such purported proofs were found but each turned out to contain an error. The more time passed, the harder the task seemed to grow. Ultimately it became one of the famous unsolved problems of mathematics.

By the nineteenth century the situation had grown so intractable that a radically different approach was tried—and it is no exaggeration to say that this approach marked a major turning point in the history of mathematics. In the 1830's the Hungarian mathematician Bolyai and the Russian Lobatchevsky announced the result of an extraordinary venture. They decided to take seriously the repeated failures to prove the parallel postulate. They decided to entertain the notion that in reality it might be false. They replaced it with an alternative: the postulate that through a point *an infinite number of different lines* could be drawn parallel to a given line. And then they went on to see what theorems could be derived.

In 1854 the German mathematician Riemann extended their idea in the opposite direction. He constructed a geometry based on the postulate that *no* lines parallel to a given line can be drawn. The results of these two alternative theories seem strange. In Riemann's geometry closed figures can be drawn using only two straight lines. In both geometrics the sum of the interior angles of a triangle is not 180 degrees, and the circumference of a circle is not π times its diameter.

What are we to make of such a state of affairs? How can three completely different geometries exist side by side? One feels an urge to accept Euclidean geometry as the more fundamental of the three, and to relegate the Bolyai-Lobatchevsky or Riemannian as mere variants—inferior. But they are not inferior. Each one of these theories is just as rigorous, just as complete, and just as respectable as the other two. Nothing exists in all of mathematics to tell us which of the three to accept as true.

To a pure mathematician there is nothing so very strange about this. Mathematicians do not need to know whether a theory is true. They only need to know whether it is consistent. To a mathematician geometry is not so much a theory as a game, an abstract set of rules one plays by. We do not ask whether bishops "really" move along a diagonal in chess. We accept that in this game that is what a bishop does. Nor do we ask whether poker is more real than bridge. And to the mathematical mind so it is with non-Euclidean geometries.

But most people are not mathematicians. Most people want to know whether parallel lines meet—real physical lines, long straight objects that can be touched. In particular, Albert Einstein wanted to know.

It was the fundamental insight of Einstein that this is not a question for the mathematicians at all. It is a question for physicists. What mathematics has shown is that a number of different geometries are logically possible. But as to which of these is true of the natural world, that is the world's own business. It is not up to us to decide. If we want to find out we must ask the natural world. We ask it by the usual methods of science: by observation and by theory-building. Einstein built a theory.

General relativity is a theory of physical geometry. It is a geometry not of flat, two-dimensional planes, nor of three-dimensional space. It is a geometry of the full four-dimensional world of space-time. Einstein decided that it is *matter* that determines this geometry. Far from material objects it is Euclidean, but the closer one approaches an object the more this geometry is distorted. Small objects distort it only a little. Massive things, on the other hand, distort it a lot.

At this very moment all of us are in close proximity to a large, massive object. This object is the Earth itself. Space-time geometry in our vicinity is therefore non-Euclidean. But although the Earth may seem massive it is not exceedingly so—not on the scale that relativity is used to dealing with. In fact, local geometry departs only slightly from the Euclidean, which is why no experiment has ever succeeded in detecting the difference. Such experiments, however, test the geometry of only part of the space-time continuum—the three-dimensional part. To study the full continuum we must look at how objects move through space in time. And here the distortion of geometry becomes very easy to observe indeed.

We call it gravitation.

Gravitation is a distortion of space-time geometry. That is *all* it is. It is not a force. It is not an attraction. We may be used to thinking of it in these terms, and even be able to persuade ourselves that we can feel its force. But all this is an illusion. Every time I stumble and fall I am not responding to a pull at all. I am performing a geometrical act.

Einstein formulated a precise, mathematical statement of his principle that geometry is distorted by matter. This statement is his gravitational field equations. When you solve these equations what you have found is the geometry. Karl Schwarzschild did this. His solution is the black hole.

The geometry of a black hole is not Euclidean; nor is it that of Bolyai-Lobatchevsky or Riemann. It is something completely different—a geometry no one had even thought of till Schwarzschild found it.

The many strange effects produced by the hole are all consequences of this geometry and only in geometrical terms can they be properly understood. Consider as an example the helpless fall that awaits any object that enters the hole. Chapter 10 ascribed this to an infinite force of attraction within the horizon. But such an explanation uses improper concepts: notions of space and time as separate, and of gravitation as a force. Relativity's space-time language explains it differently. It says that within the horizon geometry is so radically distorted that there does not even *exist* a path the object could follow that would carry it outward. Every direction it can head carries it straight down into the singularity.

Such a situation is inconceivable in Euclidean terms. But it is not inconceivable if we adopt a different geometry. An analogy will help here. I stand upon the surface of the Earth exactly at the north pole and I take a step. What direction am I walking?

A little thought shows that no matter which way I walk I am heading south. Figure 46 illustrates the situation. If I keep going in a straight line ultimately I will wind up at the south pole. Straight lines upon the earth—great circles—diverge outward from the north pole and yet they manage to converge upon a distant point—just as helplessly, just as inescapably as falling objects within a black hole converge upon the singularity. They converge not because of some force, but for purely geometrical reasons.

What of the bending of light by gravity? Chapter 9 sought to understand this by means of a trick: the trick of attributing mass to

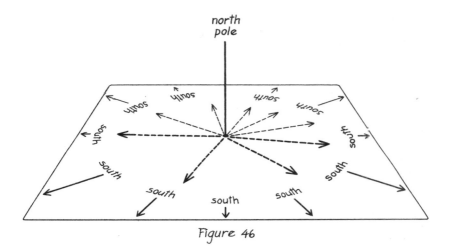

Figure 46

light and letting this mass fall. Unfortunately the attempt was un-successful for it predicted just half the true effect. It is only in the geometrical language of relativity that a proper understanding of this phenomenon can be found. Why are rays of light bent by gravity? They are not bent at all.

Here too the analogy of the spherical geometry of the Earth is helpful. Consider an airplane trip from New York to Rome. Rome is due east of New York, and one would think that to go there the plane ought to head straight toward it—due east. But this is not what the airline companies do. They send their flights on a giant arcing path: up the coast of North America, far to the north over the Atlantic, and down then in a generally southeasterly direction over Europe. As drawn in Figure 47 the path seems bent.

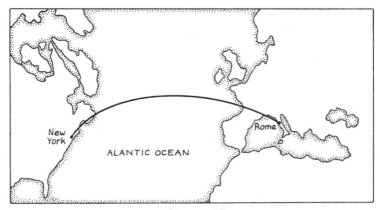

Figure 47

It is not, of course. Airlines fly the shortest distance they can. They fly the great circle route, and if this route appears curved, it only means we have been using the wrong map. It is the map that introduces the distortion. Indeed this distortion is easily visible upon every world map. Antarctica, for instance, appears as a long thin strip of land; and the south pole, a mere mathematical point, is stretched out into a line. No wonder the great circle route is distorted. And so it is with the arcing paths of light rays and the twisting corkscrew path of the Earth in space-time as it moves about the Sun. Their curvature is an illusion.

If the map is wrong, why not use a right one? The problem is that we cannot, and this for the simple reason that maps are *flat*. If we try to transfer the non-Euclidean geometry of the surface of the Earth to the Euclidean geometry of the printed page, we are bound to introduce distortions. And similarly, there is no accurate way to represent the Schwarzschild geometry of the black hole in a two-dimensional picture.

But maps are handy things for all of that. A variety of representations of black hole geometry can be sketched and so long as their limitations are kept in mind they can be valuable aids. Just as there are many maps of the Earth—Mercator projections, polyconic projections—so there are many maps of the black holes. Figure 48 shows one.

Figure 48 is a space-time diagram of the surface of a collapsing star. An instant of time in the collapse—a snapshot—is represented by a horizontal slice through the diagram. The intersection of this slice with the cone representing the star is a circle defining its surface at that instant—we must mentally add the extra dimension which changes this circle into a sphere. As time passes, the star grows smaller and eventually collapses within its Schwarzschild radius. At this instant the horizon forms. Inside the horizon the star continues collapsing and, shortly thereafter, turns into a singularity which continues forever.

Also represented in Figure 48 are the paths of light rays. Ray A was emitted from the surface of the star when it was fairly large. It is nearly straight, implying that gravity was so weak at this stage that geometry was nearly Euclidean. Ray B, emitted when the star was smaller, is more strongly curved. As for ray C, it set out the instant before the star had reached its Schwarzschild radius. So non-Euclidean is geometry at this point that it circles fully twice around the star before escaping to infinity. And finally, ray D was emitted after the

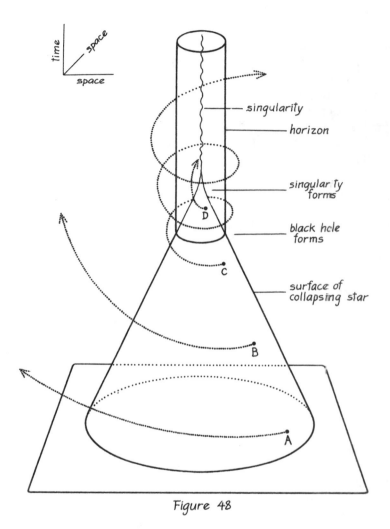

Figure 48

black hole had formed. This ray is trapped and it falls onto the singularity.

Such diagrams are useful for analyzing certain aspects of black hole physics. But they do not show everything. Figure 49 is another attempt. It is meant to suggest the *curvature* of space-time in the hole's vicinity. The flatter the surface shown in Figure 49, the more nearly Euclidean the geometry there. Far from the hole the surface is flat and geometry is Euclidean, but the closer one approaches the sharper the curvature. Finally, the shaded gray cap on the bottom of the funnel represents the star itself.

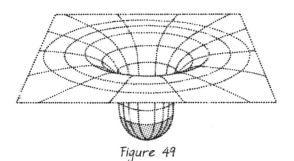

Figure 49

In contrast to Figure 48, which shows the entire history of the star's collapse, Figure 49 represents the geometry at a particular instant. Further stages in the collapse can also be illustrated. The star grows smaller, the curvature of space-time more severe (Figure 50).

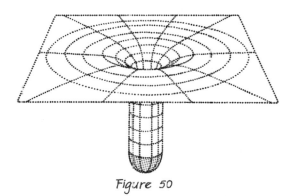

Figure 50

Ultimately the star entirely vanishes from the diagram, as in Figure 51.

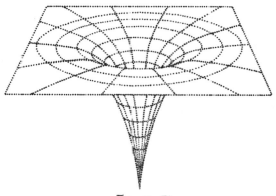

Figure 51

It has been crushed into a singularity. Now only the curvature remains.

The advantage of "maps" such as these is that they are new ways of looking at things—and the advantage of *these* is that they can lead to new ideas. The above structures suggest a new kind of geometry. They suggest a game. Take two such diagrams. Turn one upside down. Cut off the shaded gray areas representing the stars and paste the two funnels together. We get the structure shown in Figure 52.

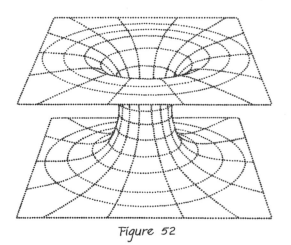

Figure 52

Of this structure the upper half is nothing new, and simply represents the outskirts of a black hole. The bottom half, though—the "upside-down hole"—is something entirely different. It is a *white hole*. The white hole, like the black hole, is an exact solution to Einstein's gravitational field equations. It is a possible geometry. It is very different from the black hole, though—in many ways just the opposite. For example, black holes trap light within them: white holes, on the other hand, emit light all the time. They shine steadily. One has no way of knowing, however, just how intense a given white hole may be. One might be exceedingly brilliant, the other so dim as to be almost invisible. Or again, nothing can leave a black hole: similarly, nothing can enter a white hole. Objects inside, though, can easily be expelled. From time to time things fly out into space. The white hole is an inexhaustible gusher of matter and energy, a source from which newly created matter spews forth into the universe. It might account for many of the mysterious emissions observed in the nuclei of galaxies and quasars.

Ultimately the entire white hole explodes. Just as a black hole (or, more properly, a frozen star) was formed at some point in the past by the process of collapse, so too the white hole will at some point in the future burst outward into an expanding cloud of matter. When will this occur? There is no way to tell: one simply waits until the bomb goes off. What will come out when it does? This too cannot be predicted in advance. The white hole could be made of anything.

Figure 52 shows more than just a white hole. It shows a black hole and a white hole joined together into some kind of composite structure. This structure is a *wormhole*: a bridge connecting two Euclidean geometries. Far from the wormhole on the top sheet one is miles distant from a black hole. Far from the wormhole on the bottom sheet one is equally far from a white hole. Between the two runs a bridge—and the existence of this bridge seems to imply that one can travel between them. The picture suggests that it might be possible to jump into a black hole, connect into a white hole, and burst out of the white hole *somewhere else.*

Nothing in all of general relativity gives a hint of where we might emerge in such a journey. The exit through the white hole might be billions of light years away from the black hole through which we entered. The wormhole might serve as a tunnel connecting two remote portions of the universe. If so it would provide a kind of shortcut between them: a way to reach the distant stars without actually traversing all the intervening spaces. Alternatively, the wormhole might connect two geometries entirely separate from each other. It might be a point of contact between two parallel universes—two spaces, each with its own stars, galaxies, and planets, each existing side by side but entirely disconnected save by this one fragile link.

The white hole and the wormhole lie at the very edge of research in relativity today. As things now stand their status is unclear. They are the subjects of fierce debate and intense research activity. Are they possible objects? Most scientists are not sure.

Much of the debate surrounding the white hole can be understood by returning to its very obvious relation to the black hole. In many ways a white hole is just the opposite of a black hole. In fact this opposition is more than skin deep: a white hole is a *time-reversed* black hole.

What is time reversal? Its nature can be understood with the help of an example. The example is a movie of a pond in a meadow.

As the movie begins, faint ripples are seen upon the surface. These ripples are perfectly circular in form, and they are propagating inward toward the center of the pond from its edges. They grow stronger. Water lilies on the pond's edge begin to rock to and fro, fanning the waves still higher. Water at the very center of the pond begins to bob up and down. The pattern of waves—perfectly circular, converging exactly to a point—tosses this water ever more severely. Now a commotion begins beneath the surface. A jet of water thrusts upward. Carried along by the force of this jet, an acorn is tossed to the sky. As it bursts through the surface of the pond it strikes against the converging pattern of waves in just such a manner as to cancel them out. The ripples cease in an instant. The acorn soars upward, a bird appears flying backward across the pond, and it catches the acorn in its beak.

The movie, of course, was a game. Someone filmed a bird as it dropped an acorn into a pond, and then played this movie backward. What the backward movie showed was the time reversal of the original process. Similarly, every act of a white hole is the time reversal of the corresponding act of a black hole.

But strange as it may sound, what the time-reversed movie showed is physically possible. It described something that could actually occur in the real world. Such a sequence of events could be produced as follows. Begin by placing tiny paddles around the outskirts of the pond. At a signal these paddles begin agitating the surface in unison. They create waves propagating inward, and they are adjusted so that the ripples converge precisely to a point. Turn up the electricity: the waves grow stronger. Now dive beneath the surface of the pond and fan tiny currents in the water. Create a jet. The jet picks up the acorn lying on the bottom and thrusts it upward. Rise into the air and alter the very muscles of the bird so that it prefers to fly backward.

This procedure violates no law of nature. It can be done. But this does not mean it can be done with ease: indeed it would be quite impossible to carry out in practice. Ultimately the motion of every atom in the pond would have to be influenced. This is why such scenes are never found in nature. They are very difficult to arrange.

Many workers in the field believe that the white hole cannot exist for the same reason. The white hole is possible but it must be arranged, and it is hard to see what could do the arranging. Something would have to push matter into a singularity and create the necessary geometry. Something would have to arrange for the singularity to spit things out from time to time. Something would have to determine how much light the singularity emits, and just when it would finally explode outward. And many scientists believe that such arrangements are simply impossible to perform.

On the other hand, it is the *singularity* that must be influenced in this way—and singularities obey their own rules. They do not obey the ordinary laws of physics. Indeed they are places where ordinary physics comes to an end. Who can say? Perhaps this is what a singularity is: a place where nature conspires to produce just those arrangements we find so strange. Only when physics finally encompasses the singularity will we know for sure. Only then will we know whether white holes are possible.

What of travel through the wormhole? If it does develop that white holes exist, can the wormhole be used as a shortcut to the stars? The first thing to be said is that this would not exactly be a rapid means of transportation. If I were to jump into a black hole en route through a wormhole, it would take forever for me to enter the hole, so far as the distant universe was concerned, and not until an infinite amount of time had passed would I emerge on the other side. No matter how distant my exit point it would have been far quicker to walk. So far as I myself am concerned, of course, the trip would be very rapid—but I would emerge into a world infinitely far removed in the future.

There is a further problem. Diagrams meant to suggest the geometry of space-time represent its configuration only at an instant. As time passes the geometry evolves. In the case of a black hole the star producing the hole collapses inward and shortly reaches a singularity. A wormhole suffers a similar fate. Initially it might be very large but it rapidly contracts. In a short amount of time it has contracted all the way. The tunnel pinches off, as shown in Figure 53.

The implication is that a wormhole is a temporary affair. The avenue of communication it provides between two regions of space remains open only for a while. In order to get through it one must travel very rapidly, and failing this one will be caught in the pinch and engulfed in a singularity. Automobile drivers slow down when approaching a dangerous passage. Travelers to a parallel universe speed up.

Unfortunately it develops that one must travel *faster than a ray of light* in order to pass safely through a wormhole—and according to relativity this is impossible. Within the framework of Einstein's theory neither a material object nor a wave can move so fast. Astronauts on board a rocket, light rays, radio waves . . . all are caught in the collapse and crushed into a singularity. The tunnel cannot be used.

A close look at relativity, however, suggests a possible exception to this conclusion. The theory does not exactly claim that nothing can go faster than light. It claims that nothing can accelerate past this velocity. It places a limit on the velocity which an initially slowly

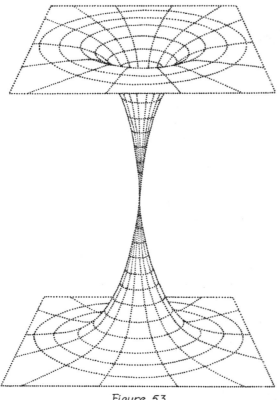

Figure 53

moving object can attain. This limit arises because as an object's velocity increases its mass does too. The object becomes ever harder to accelerate; just at the speed of light its mass is infinite and no force, no matter how enormous, would be capable of accelerating it still more.

But what of an object already traveling faster than light? Here the claims of relativity are just the reverse. The speed of such an object grows harder and harder to change as it *slows down*. Ultimately, when traveling just faster than a ray of light, its speed becomes quite impossible to affect. Such an object would be doomed to continue traveling forever, incessantly hurtling along at speeds in excess of that of light. And such an object could easily pass through a wormhole.

Such hypothetical objects have been named *tachyons*—from *tachys*, the Greek word for "swift." It is very difficult to decide whether they could actually exist. No experiment has ever succeeded in detecting one. Furthermore, their predicted properties are so strange that one is tempted to say they could not possibly be real. For example, as the energy of a tachyon decreases, its velocity increases. Ordinary objects lose energy and come to rest: tachyons lose energy and speed up. The mass of a tachyon is imaginary—the square root of a negative number. And finally, the tachyon destroys any possibility of a unique ordering of events in time. Within our experience it is always possible to decide whether one event occurred *before* or *after* another. This remains true in ordinary physics as well. But if the tachyon exists, such a distinction is no longer possible. If a gun could be constructed that shot bullets faster than light, the most fundamental notions of time would be upset. You could shoot me with such a gun—and a third party would determine that it was my body that rose up from death, and emitted a faster-than-light bullet with such accuracy that it subsequently lodged in your gun.

Modern research on black holes has moved far beyond Schwarzschild's original discovery, and nowadays it is couched in abstract, speculative terms. Scientists study the mathematical properties of new geometries, and the issues they address sound positively bizarre to the untutored ear. Rotating black holes possess features that put more ordinary holes to shame. By dipping again and again into a spinning wormhole, a traveler could pass from one to a second universe, and then on to a third, and a fourth, and so on endlessly. By a suitable choice of path he could travel backward in time. He could watch himself being born. He could shoot himself just before entering the rotating wormhole—in which case he never entered that wormhole at all, and never did shoot himself. . . . Philosophical questions on the nature of cause and effect in physics come to the fore in debating whether such geometries should be taken seriously. And if an electric charge is added to the hole its properties grow stranger still.

Perhaps most extraordinary of all is that these things flow from a theory proposed so very long ago. The age in which Albert Einstein developed relativity is long gone now, and few of its concerns remain alive to grip us. Nevertheless this creation, this marvelous structure, still rises to amaze and excite us. Nowhere in all of physics does general relativity reach so complete a development as it does in the study of black holes. Nowhere else does its beauty shine so brightly. The power of this theory, its inexhaustible richness, still hold us in its thrall.

PART THREE

———•———

THE CONTEXT

12

The Chandrasekhar Limit

Gravitation is the most important physical principle in astronomy. It dominates everything. Neutron stars and black holes are examples of what it can do. They are what happens when gravity wins.

But gravity does not win all the time. It affects ordinary stars such as the Sun, but they are able to overcome it. They resist its compression and remain extended into large, diffuse structures. These stars manage to achieve what pulsars and black holes cannot by being hot. Because they are hot they develop an enormous pressure in their interiors, a pressure sufficient to oppose gravity.

But why are stars hot? A hot brick placed in the absolute zero of space quickly cools; the Sun has kept its temperature high for more than four billion years now. The difference is that a star is a furnace. It is burning fuel—nuclear fuel. The Sun's high temperature is maintained by the energy released from the conversion of hydrogen into helium: a controlled hydrogen bomb.

No fuel lasts forever. Eventually the Sun is going to run out. So will every other star in the sky. An ordinary furnace simply cools off once its energy supply is exhausted, but stars depend for their very existence on their fuel. Once it is gone they are in jeopardy. This will occur to the Sun about six billion years from now—comfortably far away. Other stars, though, are closer to the end. Many have already reached it. What then? After a star has consumed the last dregs of its nuclear energy reserves, what will it do?

It will contract. Once the star begins to cool, its internal pressure drops, and once this happens gravity has won. For billions of years the two were in balance; now nothing remains to oppose gravitation. The star shrinks—perhaps until it has become a pulsar or a black hole. *Pulsars and black holes are part of the same subject. They are what happens to a star when it runs out of nuclear fuel.*

This final collapse is inevitable. Every star in the sky will undergo it. As for its details, though, we are much in doubt. There are many possibilities. The star might slowly shrink, taking millions of years to ease into its final state. This will probably occur to the Sun and stars like it. A second possibility was described in Chapter 7: catastropic implosion to a black hole. To make things vivid, that description was couched in terms of the Sun, but there is little chance of this actually happening to it. It may well occur, however, to other stars. A third possibility is the implosion-explosion case described in Chapter 1: a neutron star is formed from the core of a star, while the outskirts explode away as a supernova.

We do not understand very well why in some cases a neutron star is formed and in others a black hole. The difficulty is the formidable complexity of the collapse process. It is a violent, messy event. The basic principles of physics are all in hand, the techniques whereby computers are used to explore their consequences well known. But present-day computers are neither big enough nor fast enough to handle the problem. It takes a more powerful machine. When the next generation of computers is available, our understanding should be much improved.

Scientists often refer to the steady shining of a star as its ordinary lifetime, and the exhaustion of nuclear fuel as its death. But the metaphor is not a good one. Corpses are not usually so active and full of surprises as pulsars and black holes. It is better to think in terms of resurrection, or death and transfiguration. I myself think of the collapse process as a metamorphosis. A collection of matter spends billions of years shining as a star. When this phase of its existence is over, the star undergoes a transformation. Something new and beautiful rises on golden wings.

New and beautiful but also strange—incredible. There is a tendency when thinking about pulsars and black holes to lapse into disbelief. It is hard to imagine they could be real. Scientists too are not immune from this reaction. No matter how far experiment and the laws of physics have carried one, it is hard to take the next step. One naturally looks around for a way out, some method of staving off gravitational collapse.

There is a way out: *degeneracy pressure.*

Degeneracy pressure is a quantum-mechanical phenomenon: it arises from the Heisenberg uncertainty principle. This principle states that it is impossible to know exactly where a particle is and how fast it is moving. Ordinary physics—the physics of the nineteenth century—had shown that at absolute zero every particle in a gas would be holding still, and so would exert no pressure. But according to Heisenberg it is never possible to claim that a collection of particles is exactly at rest. There is always an uncertainty, some residual motion left over when all that due to temperature is removed. The pressure arising from this motion is degeneracy pressure.

The important thing about degeneracy pressure is there is no way to make it go away. Reducing a star's temperature does not reduce its degeneracy pressure. Even at absolute zero it persists. And if this pressure is enough to counteract gravity, the star will remain in equilibrium forever. It need not collapse to become a pulsar or a black hole. Furthermore its new configuration is permanent: fuel lasts only so long, but degeneracy pressure continues forever.

There is a class of star which has managed to reach this final haven: the white dwarf. White dwarfs are stars that have exhausted all nuclear energy reserves and contracted—smoothly and steadily—to the point that degeneracy pressure balances gravity. From now on they are safe from catastrophic collapse. They are a way out.

But this way out is not available to every star. For some there is no escape from cataclysmic implosion to become either a neutron star or a black hole. Scientists resisted this conclusion for many years. The man who dragged them to it, kicking and screaming all the way, was Subrahmanyan Chandrasekhar.

Chandrasekhar was born in the city of Lahore in what is now Pakistan in 1910—coincidentally, the year in which the first white dwarf was discovered. When he was eight his family moved to Madras in the southern part of India. Even as a child his life was steeped in science. His uncle was the famous Indian physicist Raman, who won the Nobel Prize for his discovery of the Raman effect. "This made a big impact on me," Chandrasekhar recalled in an interview conducted by the American Institute of Physics, and "the atmosphere of science was always at home. But I would say that my really serious interest in the kind of things I was later to do originated only when I was in college."

In college he was an unusual student. He excelled in mathematics and physics. He read everything he could get his hands on. "I read

Sir Arthur Eddington's *Internal Constitution of the Stars*. I read Arnold Sommerfeld's *Atomic Structure and Spectral Lines*, and Arthur Compton's *X-rays and Electrons*." He read scientific papers in the library. These were the marvelous, heady days of the creation of quantum mechanics: thousands of miles away in Europe, Heisenberg was formulating his uncertainty principle, Bohr his theory of atomic structure, and Schrodinger the wave equation. Back in India only faint echoes penetrated. Chandrasekhar strained to hear. Before long he knew more about the new physics than his teachers did; and what he learned, he learned by himself. "I wasn't taught quantum mechanics in school," he has said. "I learned it from Sommerfeld's *Atomic Structure and Spectral Lines*, and Sommerfeld's book is one from which one could read and learn oneself. . . . [It] is really quite marvelous, and anyone with an interest in science could read it, and verify every single step, and understand it. So is Compton's book.

"Sommerfeld visited India in 1928 and I went and talked to him quite a bit. Of course it was terribly bold of an undergraduate student to go and talk to the great man. But I had read his book by myself, and had thought this was the be-all and end-all of physics. So when I went to him I told him proudly that I had read his *Atomic Structure and Spectral Lines*. He promptly told me that physics had changed considerably since he had written it. He told me about Schrodinger's wave mechanics. It was the first I had heard of it."

Sommerfeld gave the enthusiastic young student copies of his papers on the new quantum theory. Chandrasekhar read them. He read other papers. He did some original work of his own, and published two research papers while still an undergraduate. His teachers could not understand them.

"There was a competition at college to write an essay on the quantum theory, and I could easily do this because I had studied Sommerfeld's and Compton's books with great enthusiasm." Chandrasekhar won and the prize was a book. "I was asked whether I wanted any particular book. And I said yes, I would like to get Eddington's *Internal Constitution of the Stars* because I had seen it in the library. Of course, it was written in a marvelous language and the early chapters are very easy to read, even for someone whose knowledge was as inadequate as mine was. It was a book I could start reading and go through." And so he learned the new astronomy as well.

In 1930 Chandrasekhar graduated from college. On the strength of what he had published, he won a Government of India Scholarship to do graduate study at Cambridge University in England. He was twenty years old and he was on the verge of an extraordinary discovery.

The pressure in a star does more than just support it against gravity. It determines a number of the star's most important features. It sets its size—by puffing it up. It sets the rate of nuclear reactions in its core—for these reactions are sensitive to it. It sets the brightness of the star—for the radiated energy is produced by these reactions.

This pressure cannot be directly measured, for conditions in the center of a star are hidden from view by hundreds of thousands of miles of white-hot gas. But it can be predicted. The methods are those of mathematical physics: one sets down the physical principles governing the star, translates each into an equation, and then solves the equations. The solution tells much about the star: its size and brightness, the pressure, temperature, and rate of nuclear reactions everywhere within. By 1930 a sizable body of information had been produced in this way. The techniques were well in hand. Chandrasekhar knew them—he had learned them from Eddington's book.

Nowadays an Indian student on his way to England would travel there by plane, and he would arrive within a day. Chandrasekhar went by boat. The trip took time, and while underway he set himself a problem. He decided to analyze the structure of white dwarf stars using these techniques. He wanted to concentrate on the newly discovered phenomenon of degeneracy pressure—to see what effect it had on these stars. Quantum mechanics was brand new in 1930 and no one had ever done it before.

It was a perfectly straightforward mathematical problem. But it led to an exceedingly strange result.

Even before he arrived in England Chandrasekhar had solved his equations—or partially done so. For white dwarf stars of low mass the solution he found made sense. But he could not solve them for high-mass stars. More than that: he was able to prove that in this case *there was no solution.*

An equation without a solution is a question without an answer. It was not that the equations governing high-mass white dwarfs were hard to deal with. They were impossible to deal with. For example, the equation $x = \text{cosine } x$ can only be solved on a computer—but we are assured the solution exists. We know there is a number equal to its cosine. But the equation $x^2 = -4$ has no solution at all. No matter what number you choose, so long as it is real, squaring it will never produce minus four. A computer would hang up on this problem forever.

Chandrasekhar was facing the mathematical equivalent of a Zen koan.

When confronting an insoluble problem it does no good to beat one's head against it. The smart thing is to sneak around. Once he had convinced himself that his equations were insoluble Chandrasekhar changed the question. He asked why they were insoluble. What was the difference between low-mass and high-mass white dwarf stars?

He did not solve that problem on the boat. It took years of effort in England to sort the matter out. The final answer he obtained can be understood by means of an experiment. Begin with a low-mass white dwarf—a well-behaved one—and increase its mass step by step. Do this by the simple expedient of adding matter to it. Asteroids, planets—dump them in. Each is absorbed into the white dwarf and amalgamated into its structure. What happens to the star in the process?

As it grows more massive a subtle change takes place in the interplay of forces within. Gravity is balanced by degeneracy pressure at every step, but as the mass grows higher the balance shifts to an ever more precarious one. Eventually it switches from a *stable* to an *unstable* equilibrium. At this critical mass the star is like a pencil balancing on its tip. The tiniest of shudders tips it sideways. . . .

And the white dwarf collapses inward.

The heaviest possible white dwarf has a mass 1.4 times that of the Sun. This number is referred to as the Chandrasekhar limit. Any star less massive—the Sun, for instance—will smoothly evolve into a white dwarf once it has exhausted its nuclear energy reserves. But a more massive star cannot do this. Once out of fuel it may try to become a white dwarf but that configuration is unstable. It implodes —catastrophically. In a matter of seconds it is transmuted into a pulsar or a black hole. Many of the bright stars that make up the visible constellations will suffer this fate. Sirius, Vega, Rigel—they will ultimately collapse. Future generations will detect them not as stars but as sources of bursting radio emission, or as distortions in the fabric of space-time.

As the mass of a white dwarf is increased, edging it closer and closer to the Chandrasekhar limit, the individual particles within it move more rapidly. At the limit they are traveling at nearly the velocity of light. Under these circumstances relativity comes into play—and it is this *relativistic degeneracy* that is responsible for the instability Chandrasekhar found. The impossibility of high-mass white dwarfs flows from the subtle and delicate interplay between the equations of stellar structure, Einstein's theory of relativity, and the uncertainty principle of quantum mechanics. Three great triumphs of twentieth-century physics unite in forcing upon us the inescapability of gravitational collapse.

But in 1930 almost all of this lay in the future. Neutron stars had not been invented yet, and the Schwarzschild solution was hardly common knowledge. As he got off the boat in England, Chandrasekhar had the results of some calculations packed away in his suitcase, and in his head the first faint glimmerings of an extraordinary vision. He was twenty years old, from the colonies, and he was on his way to one of the intellectual centers of the world.

He was not happy there at first: "Getting to England was a shattering experience. To suddenly find myself in an environment where there were people like Dirac and Eddington and Rutherford and Hardy, not to mention all the other well-known names; it was a very strong sobering experience. I had been extremely optimistic in India, but once I came to England I became very sobered, if not humiliated. I didn't really know whether there was any possibility for me to accomplish anything in the world in which I found myself."

R. H. Fowler was one of the scientists whose papers Chandrasekhar had read while a student in India. "I remember meeting Fowler for the first time when I arrived in England. I met him in his rooms in Trinity College. He had asked me to come and see him. I gave him the manuscript of my paper on the white dwarf mass limit. I didn't understand at the time what this limit meant, and I didn't know how it would end. But it is very curious that Fowler did not think the result very important.

"Fowler did not have an office in those days. He used to meet his students in the library of the old Cavendish Laboratory. And I used to stand outside the library, sometimes for an hour or two, hoping I would get a chance to see Fowler. Most often I did not. Somehow or other I felt I didn't belong there. It seemed to me that there were far too many big people, far too many people doing important things, and that what I was doing was insignificant by comparison. I suppose I was afraid. Even now I remember exactly how I felt, standing there waiting for Fowler. . . . I saw him once in six months."

As he had in India, Chandrasekhar worked alone. The difference was that here he was a foreigner, and overawed, and at times he was very lonely. He had a room of his own in which he worked, and he only met people when he went out to lectures or to meals. "I didn't mix with people very well. I felt shattered in their presence, and I essentially recoiled within myself. After I had been two years in Cambridge, and done all this work, I hadn't made any impression on the environment to the extent that I could judge myself. I was just by myself. And I didn't know whether I was making any headway or not."

As he worked out the theory of white dwarf stars, Chandrasekhar could not even be sure the task he had set himself was worth the effort. Would it lead to a significant increase in understanding? The thought that he had stumbled upon an important discovery "occurred to me several times. But I kept away from it. Because, somehow, the fact that this was going to play a very fundamental role —I was not willing to draw that conclusion. I was, in a sense, too diffident to draw such a conclusion, even though the thought insistently occurred to me."

In this, at least, he was not alone. It was his way of dealing with the question every scientist must face, and not just once but over and over again: whether the research one is doing adds up. Usually there is no way to answer this question short of committing the years until the result is obtained and seeing what it looks like. More often than not the decision to begin research on some problem is a decision to take a risk, to gamble time. Many are the projects that simply fail and are abandoned, foundered on some difficulty too severe to be overcome. Many also are those carried through to a successful conclusion only to find the answer hardly worth the knowing. Worst of all is to spend the years and by accident, by bad luck, not to ask that crucial question which opens the door to an important discovery. Research is often a matter of chance, and there are many fine scientists who have never had the good fortune to stumble upon the Lucky Break that makes their reputations. Conversely, plenty of Nobel Prizes are awarded for luck. In the long run it turned out that Chandrasekhar was lucky in his choice of problem. But he did not know this at the time.

"In 1933, after I had finished my work for the Ph.D., I went to ask Fowler whether I had any further scope in England. He said, 'Well, you can apply for a fellowship in Trinity College, but I don't think you have much of a chance.' I applied anyway, but I was so sure that I would not get the fellowship that I arranged to leave Cambridge on the day it was to be announced. And on the way to the station I stopped at the College to find out who were the people that had got elected. I was astonished to find my name among the people who were elected. And I remember telling myself, 'Well, this has changed my life.'

"I had suddenly become a part of 'Cambridge.' I could sit now at the same table with all the others. Gradually I could find people with whom I could talk, discuss, and indeed become friends."

After a number of years Chandrasekhar reached his conclusion. He decided that high-mass white dwarfs could not exist. But there were other scientists who were equally sure that they could.

The physicist E. A. Milne, a colleague and a personal friend, wrote to him that "it is clear that matter cannot behave as you predict." Sir Arthur Eddington spoke of his work as follows. A white dwarf of more than 1.4 solar masses . . . "apparently has to go on radiating and radiating and contracting and contracting until, I suppose, it gets down to a few kilometers radius where gravity becomes strong enough to hold the radiation and the star can at last find peace. . . . I felt driven to the conclusion that this was almost a reductio ad absurdum of the relativistic degeneracy formula. Various accidents may intervene to save the star, but I want more protection than that. I think that there should be a law of nature to prevent the star from behaving in this absurd way."

Milne never got around to explaining to Chandrasekhar what made him so sure, nor just where his friend had gone astray. Eddington spoke in terms of absurdity—hardly an astronomical concept. An argument was in the offing, and it was due to be couched in completely unscientific terms.

Intuition and taste have an important role to play in science. Chandrasekhar had solved some equations. Now it had to be decided whether his result was reasonable—whether it made any sense. And if it made no sense people were not going to throw out his solution. They would throw out the equations he had solved themselves.

Everything he had done was based on quantum mechanics and its uncertainty principle, and quantum mechanics was brand new in those days. By now it is hallowed, enshrined by decades of research and confirmation, but in the thirties it was largely untried. If the new theory led to an unacceptable result, one simply ignored the result. But how to decide what was acceptable?

In judging the "reasonableness" of a conclusion there are no hard and fast guidelines—none whatever. It is a different business from arriving at the conclusion in the first place. To do *this* one performs experiments, solves equations, and the like. It may be hard, it may be set about with pitfalls for the unwary—but at least it is well defined. There are rules for judging the formal validity of a result. But beyond this point ambiguity reigns. If the rules are straightforward for deciding whether something is mathematically correct, those for judging whether it makes any sense can hardly be said to exist. And yet in this gray area much of the actual progress of science occurs.

Any competent scientist develops an intuitive feeling for the field, a nose that allows him to smell things out. Good scientists have good noses, and they have an uncanny ability to sense the truth of things. Over and over again people allow themselves to be guided by these intuitions. Over and over again a scientist will declare he does not "believe" some result—and then he will proceed to ignore it. Another will declare he has no "faith" in someone, and he will be prepared to resolutely discount everything this colleague says. These people have smelled a rat, and they are resolved to save themselves years of toil by avoiding a blind alley. A third will declare, often on the basis of utterly insufficient evidence, that something "must be true," and he will then act as if it were. This man's nose has ferreted out a fruitful area of research, and long before any rational evidence can be set forth, he has embarked upon it.

Such terms as "belief," "faith," and the like may sound strange. But they should not. The impression that science is solely guided by logic rests on a lack of appreciation of the layers upon layers of ambiguity that actually surround the profession.

The most telling illustration of the way in which intuition guides a scientist astray can be found in a paper by the Russian physicist Landau, written two years after Chandrasekhar discovered his limiting mass. In this paper Landau makes the same discovery, apparently unaware of Chandrasekhar's earlier work. But he gives it a completely different status: rather than taking it seriously, he uses it as proof that a new theory of physics is required. After demonstrating that equilibrium is possible only for low masses, he goes on to write that for greater masses "there exists in the whole quantum theory no cause preventing the system from collapsing to a point. As in reality such masses exist quietly as stars and do not show any such ridiculous tendencies, we must conclude that all stars heavier than [the limiting mass] certainly *possess regions in which the laws of quantum mechanics are violated.*" [Emphasis added.] Like Eddington and Milne, Landau rejected the conclusion. But in this case the conclusion was his own.

It would be easy enough to dismiss these errors by supposing those who disagreed with Chandrasekhar to be inferior. But they were not inferior. Eddington and Landau were among the great scientists of our century. Landau won a Nobel Prize; many would argue that Eddington should have too. Landau invented the idea of the neutron star; Eddington brought relativity to England and verified the bending of light by gravity. No, the real moral is far more ambiguous. It is that there is no formula for success in science—none at all. You do the work and make your choice. Then you lay your neck out on the line . . . and wait to see what happens.

During the thirties, Sir Arthur Eddington's "position in astronomy was dominant," Chandrasekhar has said. "I don't think there was any doubt in anybody's mind that Eddington was always right. There was a meeting at which Eddington and I disagreed in January of 1934. I gave a paper. Then Eddington came up and said, 'The paper which has just been presented is all wrong.' He then made a lot of jokes and at the end of the meeting everybody came and said to me, 'Too bad, too bad.' The other astronomers were certain that my work was wrong simply because Eddington had said so."

The first white dwarf had been discovered by the American astronomer Henry Norris Russell. "The following incident is illustrative of Russell's attitude. At the International Astronomical Union meeting of 1935, Eddington was President of the Commission on the Internal Constitution of Stars. Russell was the secretary and presiding. With Russell presiding, Eddington gave an hour's talk, criticizing my work extensively and making it into a joke. I sent a note to Russell, telling him that I would wish to reply. Russell sent back a note saying 'I prefer that you don't. And so I had no chance even to reply, and had to accept the pitiful glances of the audience.

"[In the early stages] I had not worked out a complete theory of white dwarfs, in which I used the exact equations. I did that for the first time in the fall of 1934. And during the time I was working on this, I was a fellow of Trinity College, and Eddington used to come to my rooms every so often after dinner to see how my calculations were progressing. He was very interested and anxious to know. And then I was scheduled to give an account of this work at a meeting of the Royal Astronomical Society. When I got the program I noticed that after my paper Eddington was scheduled to give a paper on relativistic degeneracy. And I was really annoyed, because here was Eddington coming and talking to me week after week about my work while he was writing a paper himself and he never even told me about it.

"[The night before the meeting] I went to dinner and Eddington was there. I was still annoyed because he had never told me. After dinner I didn't try to go see him. But he came up to me. And even then he wouldn't tell me.

"Then at the meeting I gave my paper. Eddington got up soon after that and said, 'I do not know if I shall leave this meeting alive, because the paper which you have just heard, the foundations of it are completely wrong.' And then he went on to make some remarks which . . . Well, if you read the published report of the meeting, you will find 'laughter' interspersed in many places."

By the end of the decade, the weight of opinion was beginning to shift somewhat. But Eddington still would not be swayed. In 1938 at an international conference, "Eddington and I really talked to each other in strong language. At the discussion an astronomer asked Eddington 'Well, Professor Eddington, there are two theories of white dwarfs. How can an observational astronomer distinguish them?' And Eddington said, 'There are not two theories.' At this I got really angry. I got up and said, 'Well, Eddington, how can you say that there are not two theories? Because you and I were in Cambridge just the other day, in a discussion with the physicists Dirac and Peierls and Price, and all three did not agree with your work on degeneracy. And to the extent that these three distinguished physicists think that my formula is right, an observational astronomer must conclude that there are two theories.'

"At this point Russell got up and said, 'The discussion is closed.' And that was the last of that.

"At the last part of the meeting, there was a big reception and lunch at the City Hall. All the greats were there. They were all up at the high table and I was way off in the corner somewhere. At the end of the meeting I was standing by myself when quite suddenly I found Eddington next to me. He said, 'I hope I did not hurt you this morning.'

"I asked him, 'You haven't changed your mind, have you?'

"Eddington said 'No.'

"And I asked, 'Then what are you sorry about?'

"Eddington just looked at me and walked away. That was my last conversation with him.

"In many ways, thinking back over those times, I am astonished that I was never completely crushed," Chandrasekhar went on. He was in his mid-twenties at the time and Eddington a magisterial, dominating figure. "Finally, in 1938, I decided that there was no good in my fighting all the time, claiming that I was right and that the others were all wrong. I would write a book. I would state my views. And then I would leave the subject."

All this happened long ago, and history has shown that Chandrasekhar was right. By now *he* is one of the Grand Old Men of Science, and he has attained an almost mythical stature in the profession receiving the Nobel Prize in 1983. As he had vowed he wrote a book, summarizing his views on stellar structure, and this book exerted a dominating influence on the development of the field, defining the subject for years thereafter. He then withdrew and turned his attention to stellar dynamics. Within three years he published a book on it, and so he went, moving from research topic to research topic, in each case finishing by producing a book so comprehensive and rigorous as to be a classic in its field. Few scientists have worked in as many different areas as he, and few have exerted so significant an influence on each. His style has been to construct a systematic body of results in each area upon which one can rely with confidence. In this he has been perhaps the last of the great systematicists. So successful has he been that by now the Chandrasekhar style has become almost as well known as the Chandrasekhar limit. There has even grown up over the years a minor industry of affectionate satire, in which papers by "S. Candlestickmaker" are submitted from time to time to the journals for publication. Each of these papers lightly spoofs the comprehensive, magisterial tone Chandrasekhar has made his own, and each is testimony to the influence he now exerts.

Chandrasekhar is also one of the great civilizing forces in astrophysics, and he has written and spoken eloquently about the cultural worth of science. "[When I was an undergraduate] it was part of the patriotism of those times to try and see what Indians could accomplish with respect to the external world. Accomplishment in science was one way of expressing what Indians could do. Patriotism is a word which is not very popular to use these days; but patriotism as it was understood in India in the twenties was expressed in everyone's wish to show that Indians could be accomplished in a way which the outside world could recognize. To accomplish in science, to show what one could do in science, was part of my motive."

But as he grew older his attitude changed to include a larger component of awe. "Why is it that what the human mind conceives as beautiful finds manifestation in nature? Take the ellipses and conic sections which Apollonius wrote about. The enthusiasm with which he writes about them! The incredible properties of these curves! And he talks about the beauty of these curves; he discerns them as beautiful. Who would have known that centuries later those curves would be found to be the orbits of the planets?

"How does it happen that the human mind thinks certain abstract concepts—and thinks of them as beautiful? And why do they find replicas in nature? The Kerr metric is an example in general relativity. Kerr discovered it in trying to explore Einstein's equations. They represent the exact descriptions of [rotating] black holes in nature.

It seems to me that there are a number of instances in which what the human mind perceives as beautiful has counterparts in nature; and this to me in many ways is a very sobering thought.

"I don't understand that. Heisenberg had a marvelous phrase, 'shuddering before the beautiful.' I would say that is the kind of feeling I have about these things.

"I am aware of the usefulness of science to society and of the benefits society derives from it. But on the other hand, so much is said about the usefulness of science that I have been more concerned with the fact that people seem to completely put aside the cultural value of science. Science is a perception of the world around us. Science is a place where what you find in nature pleases you. That one can derive joy from studying and understanding science, that one can learn science the way one enjoys music or art—it seems to me people ignore these aspects. Indeed, I would feel that an appreciation of the arts in a conscious, disciplined way might help one to do science better."

In Chandrasekhar's office, across from his desk, a photograph is hung on the wall. It was made by Piero Borello and it is entitled "An Individual's View of the Individual." It shows a person halfway up a ladder, and above is some strange symmetrical structure whose nature eludes us. Borello agreed to let Chandrasekhar have a copy, but only on the condition that he explain why he wanted it. He replied, "What impressed me about your picture was the extremely striking manner in which you visually portray one's inner feeling toward one's efforts at accomplishment: one is halfway up the ladder, but the few glimmerings of structure which one sees and to which one aspires are totally inaccessible even if one were to climb to the top of the ladder. The realization of the absolute impossibility of achieving one's goals is only enhanced by the shadow giving one an even lowlier feeling of one's position."

13

Uhuru

The Chandrasekhar limit made collapse almost inevitable—but not completely so. Eddington had wanted a law of nature forbidding such nonsense. He did not get one. But he had also mentioned "various accidents" that might intervene, presumably referring to explosions that would disrupt the star before it had a chance to implode.

This was another matter. There was no simple way of ruling it out. The only way to deal with it was to use brute force: to track theoretically the evolution of a star over its billions of years of nuclear burning—and not just one star, but each of all the different stellar classes. Even a short segment of one such calculation requires hours on a modern computer. In the early days it was harder still.

By the early 1960's a number of calculations had been pushed through all the way. Some stars exploded, but most did not. It was beginning to appear that complete gravitational collapse was a realistic possibility. An evolution in attitude was under way. One paper referred to the view that accidents would invariably intervene as "no more than a superstition"; another took seriously the idea of collapse and went on to interpret quasars as examples.

The discovery of pulsars in 1967 was a punctuation mark in this evolution. It put things in a particularly vivid light. Pulsars were examples of the very implosion Eddington had so strongly resisted. And if neutron stars, then why not black holes as well?

Bit by bit the Schwarzschild solution was changing from a mathematical curiosity into hard reality. The black hole was entering the realm of the possible. But how to find one? It was a few miles across—telescopes have difficulty picking out something so small on the Moon, let alone light years away. It was black, and superimposed against the blackness of space. The distortion in the background sky produced by the gravitational lens effect might be detected, but the distortion was strong only within a few miles of the hole.

All in all it was going to be very hard. The difficulties seemed almost too great to be overcome. As it turned out they were not overcome. The first black hole was found another way.

I was sitting with Harvey Tananbaum in his office in Cambridge, Massachusetts, and he was telling me of his trip to Malindi, a small seacoast town on the shores of the Indian Ocean in Kenya. Although Malindi is something of a resort town, Tananbaum had not been impressed by the night life there. "There were three or four hotels," he told me, "there was a beach and some swimming pools. There was a disc jockey who had a stack of records which he would take around to the various hotels. On Mondays he would play through that stack from top to bottom and on Tuesdays he would play through it from bottom to top. Thursdays and Fridays he played the flip sides. On Wednesdays you could go out to the movies—they hung up a big sheet out of doors between two poles.

"It was a dirt road from my hotel in Malindi down to Base Camp," he continued, "and if it rained heavily there would be a period of a few hours in which that road was so muddy that it was completely impassable. It was narrow and full of potholes. If you were to come upon a car going in the other direction you would honk and he would honk and one of you would move halfway into the bushes to let the other pass. Often you had to stop while the local little boy with a stick would chase his cows or his goats across the road. They had the right-of-way, of course.

"Not that it seemed to make much difference. The Italian crew down there had two cars at their disposal and some way or another they managed to get involved in a head-on collision one day on that road. With the only two cars in the vicinity.

"The green mamba was pretty common in the area. Just outside Malindi were some old Arab ruins which were known to be an area in which they lived. You had to be careful out there. And no matter where you went you always had to be careful when walking under trees, for you didn't want something dropping down onto your shoulders. One night at my hotel a young Englishman brought over a puff adder—deadly dangerous—in a bag to show around. That thing suddenly made a strike at its handler and he dropped it. The snake got away, and there was quite a panic till he grabbed hold of it again and got it into its bag.

"It was hot and fairly humid. We would go swimming and even the ocean was warm, 85 degrees or so. Bugs? No, the insects weren't too bad."

He was there to launch a satellite.

The science of X-ray astronomy was created on a Monday, and it happened in the state of New Mexico. On the midsummer night of June 18, 1962, a small Aerobee rocket was launched from the White Sands Missile Range, and before it crashed to Earth a brief six minutes later, it had detected X radiation coming from the sky.

Why a rocket? To answer this question consider an analogy, and imagine a planet entirely enshrouded in clouds. We can visualize such a world as being similar to the darkest, foggiest day that we can recall. The difference is that here the cloud cover never lifts. In consequence, no one living on such a world would have ever seen the sky.

On such a world the familiar rising and setting of the Sun would be completely invisible. The passage of day into night would be marked only by a gradual dimming of the dull level of illumination into obscurity. During the day no Sun; during the night no Moon and no stars: only a uniform gray dome stretching overhead.

If we were to live on such a world we would be capable of proving that it was round—by flying around it in a jet. We could prove that it rotated—by watching a Foucault pendulum in any science museum. As time passes the pendulum slowly changes the path in which it swings, ultimately completing a full circle. Actually the path is not changing at all: it is the ground that rotates beneath it. The pendulum's behavior proves the rotation of the Earth, and were we to live on a fog-enshrouded planet we could interpret its behavior correctly there too. Very likely we would be struck by the fact that the twenty-four-hour period of rotation agreed quite nicely with the twenty-four-hour period of the day-night cycle. In this way, without ever having seen it, we would be led to postulate the existence of some external source of illumination beside which our planet slowly spun. But there would be no way of learning how far away this object was, nor how big—nor that the Sun was round.

In a similar way, by studying the rise and fall of the tides, it is possible to demonstrate the existence of the Moon. But although this is possible in principle, it is a much more chancy business: in practice we would probably have no idea whatever of the Moon. And in any event there would be no way of knowing that it was made of rock, or covered with craters.

And no one would have the slightest reason to contemplate the existence of the planets or the stars.

This hypothetical planet is not so very hypothetical after all. It is the Earth—*if* we saw not light but X rays. It is entirely possible for evolution to produce a living organism with eyes sensitive to X radiation. But they would be useless to it here, for the atmosphere of the Earth is not transparent to X rays. It blocks them out. The above account quite accurately describes the predicament such a creature would face in doing astronomy. The only way to observe X rays from celestial sources is to get above the atmosphere—to launch a rocket or a balloon, or to place a satellite into orbit.

As early as 1956 there had been hints that various objects in the sky emitted X rays, but nothing sufficiently definite to warrant publication in the journals. The 1962 rocket flight from White Sands was the first unambiguous detection of celestial X radiation, and it is fair to say that with it a new branch of astronomy was born. The experiment had been launched by four scientists from the Massachusetts Institute of Technology and a private firm, American Science and Engineering: Riccardo Giacconi, Herbert Gursky, Frank Paolini, and Bruno Rossi. Perhaps what is most remarkable about the experiment is that it completely surprised its designers. They had built it with the expectation that it would detect X radiation from the Moon, produced when the solar wind impinged upon its surface. In fact the experiment failed to find this radiation, but it did find something else: an intense and entirely unexpected shine of X rays coming from some object completely outside the solar system.

For eight years thereafter the fledgling science was an unimaginably frustrating field in which to work. These pioneers knew almost nothing about their newly discovered X-ray sources. Each rocket flight cost large sums of money and took years to design and build. Many when finally launched simply blew up on the pad. Others rose successfully, but some component of the X-ray detector might fail to operate and the results were inconclusive. These were the early days of the space program, but the X-ray astronomers were operating on a shoestring budget. They could not possibly afford the mighty giants used by NASA. The best that could be done were short flights lasting at most a matter of minutes, and one way or another all the observations had to be jammed into this brief period of time. And even these short flights were painfully few and far between— never more than a few per year.

Returning to the analogy of the cloud-covered planet, the situation in those years was as if at great cost and inordinate effort it could be arranged to produce for a brief few minutes a small gap in the overlying clouds. The first time this is achieved, perhaps, scientists would see the stars. The second time—six months later—they might see the Big Dipper and realize the stars form patterns. If some lucky soul were to bring along a pair of binoculars or a small telescope—and there is no particular reason why he should—he would be rewarded with other sights: sunspots, or the rings of Saturn. But the knowledge gained in this way would be fragmentary and incomplete, and progress would be desperately slow. And worst of all would be the intense and never-ending frustration as the clouds closed in again.

That is how it was until Uhuru came along.

Uhuru was the first satellite built to study X rays from the sky. It did not fall back to the ground five minutes after launch. It was designed to stay up above the atmosphere, continually orbiting the Earth and making observations, for a full year. Had it lasted half that time, its builders would have been overjoyed. Such was the pace of events prior to Uhuru that if one were to add up all the time that all the rockets that had ever been flown had spent above the atmosphere, one would get something like a week. Had Uhuru remained in orbit for a single week, then, it would have doubled our knowledge of the X-ray sky. But it did not survive a week. It lasted three years.

In those three years this one satellite came close to transforming astronomy. It discovered utterly new and unexpected celestial phenomena. It discovered objects the existence of which had never even been suspected—not just one or two such objects, but vast numbers of them, whole new classes of objects. It uncovered new stages in the evolution of stars. It detected vast intergalactic clouds the existence of which had been endlessly debated but never before revealed. It observed X rays from distant galaxies as well as from our own. The most wild of speculations was surpassed by the hard data that rained down from the sky. In a brief five minutes—the time required to boil an egg—Uhuru accumulated as much data as an entire rocket flight. In the next five minutes it did so again, and on and on it continued: observing the sky as the weeks stretched into months and into years. Its operations, of course, were largely automatic, and performed in the stillness of outer space. Meanwhile, down below, astronomers were running ragged as they sought to comprehend the new view of the heavens that had suddenly opened to them. Rarely in the history of science has a single instrument made such a contribution to a field. It is as if the perpetual cloud cover were to be banished at a stroke.

None of this was cheap. Uhuru cost $5,000,000. Nor was there any way to reduce this sum. Nothing could have been done to save money by cutting corners, and the reason reveals something of the nature of space research.

As it was finally assembled Uhuru weighed some 350 pounds; a fairly heavy load. How much did it cost to lift such a weight into orbit? Think in terms of energy. A convenient unit with which to work is the calorie—it is the same unit that weight watchers count so carefully.

How much energy is expended in simply picking up such a load? The answer works out to about 400 calories. If 350 pounds is too heavy for me to manage alone, I can share the work with another by putting Uhuru on a stretcher: then each of us expends 200 calories. Now carry the stretcher up a flight of stairs. Between the two of us we have expended 1,600 calories. Raise it to the top of the World Trade Center: the elevator burns up 180,000 calories in the process. Finally lift Uhuru into orbit 900 miles above the surface of the Earth. It costs 200 million calories to lift such a weight this high.

But this is only the beginning. If all we were to do is lift a load into space and then let go it would fall back down to Earth. We must keep it up there! How to keep the satellite from falling? It cannot be kept from falling. Short of placing it atop a 900-mile high tower, there is nothing that will prevent an object in space from falling. But it can be kept from falling *back down to Earth*. Once the satellite is at the required altitude you do not simply let go— you give it a sideways kick. Kick it only a bit and the satellite will land a few miles to one side of its launch site. Kick it harder and the satellite will fall at more of an angle: perhaps it will land a thousand miles from its launch site. Now kick it harder still. Now it *never* hits the ground as it falls. It falls in a looping arc all the way around the Earth. It is in orbit.

To achieve orbit the sideways velocity imparted to the satellite has to be very precisely adjusted. Too small a velocity and it falls to Earth, too large and it flies away into space. The rocket that launched Uhuru lifted it 900 miles up and then sped it to a horizontal velocity of 17,000 miles per hour. It turns out that the energy required to accelerate 350 pounds to this velocity greatly exceeds that required to lift it into space in the first place: about a *billion* calories.

This is the energy budget of the launch: 200 million calories to raise the satellite above the atmosphere plus one billion calories to accelerate it, for a total of 1,200,000,000 calories. It looks like a lot of energy. But it is not, for a calorie, in fact, is a very small unit: a single gallon of gasoline contains 30 million of them. To insert Uhuru into orbit required a quantity of energy contained in a mere 40 gallons of gas. At today's prices at the pumps this would cost some $50 or $60.

But Uhuru cost $5,000,000. How did $5,000,000 turn into $50?

We have left something out. What we have left out is the *weight of that 40 gallons of gasoline*. The rocket did not just have to lift the satellite. It also had to lift the fuel. Not only that, but in order to accelerate Uhuru it had to accelerate the fuel as well. For example, the very last step from 16,000 to 17,000 miles per hour required a certain amount of fuel, and this fuel had to be carried all the way up into space and then itself accelerated to 16,000 miles per hour—all before it could be finally used.

As an analogy imagine that I am preparing to drive across the country—but as the result of a nationwide oil shortage every gas station along the way is going to be closed. The only gas available is located right here in my home town. I will have to take it all with me. Of course my gas tank cannot carry such a load. I will need to tow along a trailer filled with gas. The trailer will be very big: it will weigh almost as much as the car! But if so then I have miscalculated how much gas I will need. My car normally gets 20 miles to the gallon, but with a load like this it only gets 15. So now I need more gas than I had originally thought, and it will weigh still more. Now, doing the calculation over again, I realize I will need two trailers. But when hauling a load like this my car gets a meager eight miles to the gallon. Back to the drawing boards again. My estimate for the gas required keeps going up and up.

So it is with launching satellites. The payloads sent into space are small, the rockets needed to lift them gigantic. By this inescapable chain of circumstances, this ever-spiraling energy budget, the cost required to launch even the smallest of satellites into orbit becomes enormous.

But there is more. Once the satellite is up in orbit we will not be able to get it down again for repairs if something goes wrong. Nor can an astronaut be lifted into orbit to make those repairs, at least not without an additional huge expense. So the satellite had better be perfect. It had better not need any repairs. A satellite must be able to withstand the violent shaking and stresses of the rocket launch. It must be able to withstand the weightlessness of orbit, the vacuum of space, the heat of sunlight unfiltered by our atmosphere, and the aching icy cold of the shadow of the Earth. A satellite must be able to control itself. It must receive commands from down below and be able to act upon them, and radio data back down to the receiving stations. If the satellite is a telescope—an X-ray telescope, for example —it must be able to control where it is pointing. It must be able to *find out* where it is pointing. And it must do these things without the slightest mechanical assistance from any outside agency.

Satellites can fail in orbit, and often they do so for the most trivial of reasons. Perhaps someone forgot to tighten a screw somewhere, or tightened it too much. Perhaps a speck of dust has jammed a relay. Tiny flakes of paint might peel off and cluster about a magnet, subtly distorting its function. Any of these problems could be solved in an instant if someone could only get up to the satellite. A swift kick might cure everything. But we cannot get there to deliver that kick. The satellite is a mere few hundred miles away, passing overhead in its orbit every 90 minutes, but as inaccessible as if it were on the far side of the Moon.

If it were to cost $50 to launch a satellite no one would worry about all this; if it were to fail we would simply launch another one. We could launch a hundred, in the certain knowledge that by the law of averages one of them was bound to work. But we cannot operate this way. Satellites are designed and built with an obsessive attention to detail. Every test that can possibly be performed is performed, and not just once but over and over again. Every component is scrutinized intensely. Everything that might go wrong is planned for. Wildly improbable chains of accidents are invented and worried over—sometimes in joint conferences, sometimes at three o'clock in the morning or in moments of depression. The engineers who build satellites are often a compulsive lot, given to a fastidious precision that would madden the rest of us. They have to be perfect.

And so again the cost is driven even higher. And so, last of all, X-ray astronomy passes beyond the realm of Small Science. Individual scientists can do optical astronomy or radio astronomy or theoretical astrophysics. Individual universities have the money to support these things. But to do X-ray astronomy you must do Big Science. And you become involved, inextricably and inescapably, in the only institution wealthy enough to pay for it, the United States government.

All this because there are no gas stations in space.

Riccardo Giacconi's office is just down the hall from Harvey Tananbaum's. It was Giacconi who conceived of the Uhuru Project, it was he who got the government funding, it was he who oversaw its construction, and it was he who presided over the glorious days when

the data rained down. He had also taken part in the 1962 rocket flight
which had ushered in the field. "As early as 1960," he told me one
day, "two years before that first successful flight, a group of us had
gotten together and written a technical report in which we tried to
foresee the possible kinds of objects an X-ray observing program
might detect. We tried to estimate the quantity of X radiation we
might receive from nearby stars, from supernova remnants, and from
the Moon. Our interest in the Moon had to do with the solar wind.
At the time there was considerable uncertainty about it. We thought
at first that when it struck the Moon's surface the wind would emit
X rays that could be detected and which would tell us something
about it. There was a group at the Air Force Cambridge Research
Laboratories on Route 128 outside of Boston that was interested in
this question and we were able to get funding from them for the
rocket flight."

I recalled that this flight had succeeded in detecting X rays, not
from the Moon, but from something entirely outside the solar system. Had those original ideas been wrong?

"Well, the Moon is a source of X rays," he replied, "but far
weaker than we had expected in those days. But the Moon was not
my motivation for doing X-ray astronomy. It was just a way we could
get funding. We knew the Air Force was interested in this problem,
so we emphasized that aspect to them. But my real motivation was
much more down to earth: I wanted to get going.

"I knew what could be done," he went on. "We had these estimates. I knew that with an X-ray detector of such-and-such a size I
could detect radiation from the Crab Nebula and Sirius and the like.
And it was a new opportunity. No one had looked at celestial X radiation before and I had a strong intuitive feeling that it would be a
worthwhile thing to do."

But none of the estimates were relevant. They didn't detect the
Moon and they didn't detect the Crab Nebula and they didn't detect
Sirius in that first flight. They detected something else. So was Giacconi mistaken in going into X-ray astronomy? Was it blind luck that
led to the discovery of the first celestial source of X rays? As I sat in
Giacconi's office, it struck me that every time a new instrument has
been invented it has discovered something its inventor had no reason
to expect. The world's first radio telescope was built to study the
Sun. It did indeed detect the Sun—but also a blaze of emission from
a wispy film of nebula in an entirely innocuous region of the sky.
Pulsars were discovered by accident when Hewish constructed a new
kind of radio telescope, and the Van Allen belts in a similar way.

In fact, however, these discoveries are no accident. If there is anything we should have learned by now it is that nature is profligate. Nature continually surprises us by exceeding our wildest expectations. Every time we have looked at it in a different way it has revealed to us new things.

But there was more in Giacconi's mind than a cosmic faith in the boundlessness of nature: "I knew I could do it better than anyone else," he told me. "I had come into the field from cosmic ray physics and I started looking at what had been done and I noticed who all the people were. They were solar physicists, atmospheric physicists and the like. The point is that I knew experimental techniques they didn't. I knew I could do clean, high-sensitivity measurements.

"After several more flights the Air Force got tired of funding us," Giacconi went on. "They liked what we were doing, they thought it was fine stuff, but they could see it didn't have anything to do with the Moon. So I had to go back and ask for support from NASA. I was worried because I was young at the time, only thirty two years old, and fairly unknown. I decided that NASA was not going to take me seriously unless I could show them I had a program of X-ray observations spread over a number of years which was a real scientific effort—a long-range program. I wrote it up in 1963 and I took it down to Washington."

Giacconi's research proposal is entitled "An Experimental Program of Extra-Solar X-ray Astronomy," and it is something of a classic. It outlines a program extending over a decade, beginning with a series of rocket flights, moving on to an orbiting satellite, then a detector to be flown on one of the Apollo missions, and culminating in a giant instrument consisting not just of a detector but an actual telescope capable of taking X-ray photographs of the sky. The extraordinary thing about Giacconi's proposal, aside from its sheer audacity, is that nearly every mission proposed was eventually built and flown. It was a prophetic document, and it set the pattern of the field for a long time to come.

"I took it down to Washington and presented it to Nancy Roman, an administrator at NASA. I gave a little lecture, three-quarters of an hour perhaps, and at the end she said she was most interested. She said she would encourage me in the satellite which I proposed. I almost fell off my chair! I hadn't imagined in my wildest dreams she would take me that seriously." Giacconi had wanted some rockets, but Roman was taking him at his word and accepting his far more ambitious proposal for a satellite.

It is a long way from encouragement from a middle-level administrator to final approval and funding of a major program. Giacconi had to go back and write a more detailed proposal concentrating on the satellite. He did this in April of 1964 and it too was received favorably. But it was not until two years later that the final proposal was submitted, and not until a year after that that the Uhuru Project was formally underway. Then it took three years to build the instrument.

"Uhuru would have been exceedingly difficult, if not impossible, without the support of Nancy Roman," Giacconi told me. It was not that she alone had the final say in deciding whether to support the project. But "middle-level administrators in NASA have a truly enormous amount of discretionary power in deciding whether to introduce new projects up the ladder of the hierarchy. For this reason personal relationships with them are incredibly important." NASA operates in two different ways in deciding which scientific projects to support. If the agency has some particular program in mind, it sends out announcements and invites bids. For example, it may have an ongoing program of ultraviolet studies of the Sun from space, and will circulate a message that in so-and-so many years a satellite will be launched in which several ultraviolet telescopes can be accommodated. Scientists are invited to propose designs. The procedure is very much like the circulating of bids to contractors for the building of a bridge.

But this was not the position Giacconi and his group found themselves in. They were proposing an experiment entirely outside any of NASA's existing programs. Proposals like these are treated differently. They must be so impressive as to persuade NASA actually to change its direction, and anyone who has tried to alter the course of an agency of the United States government knows what this can be like. In this task it is invaluable to have the support of someone within the agency, an administrator of some power who is interested in the project.

"An important thing here is *plurality of access*," Giacconi said. "NASA has many centers and they are moderately independent of each other. Each is hustling to do its own stuff. They all have their own pet areas of interest. So you have many openings: if one administrator tells you no, then you aren't necessarily finished—you can try elsewhere. Or if NASA turns you down you can go to the National Science Foundation, or in those days to the Air Force."

Giacconi is Italian, and he did not come to this country until after he had completed his training. "I could never have done this in Italy," he told me. "In Italy there is much more resistance to new ideas." In most European countries the scientific community is organized in a very hierarchical fashion compared to ours. At the top of the heap in Europe there is a small number of very senior scientists who wield great power and direct the scientific enterprise. In contrast, the American system is much more democratic—or anarchic, depending on one's point of view. "In Italy the only way you could have done something like Uhuru would be if some very senior professor, maybe eighty years old, were to get together with a committee of other very senior professors and decide it was a good thing. Someone my age would have had no chance. And typically, just because this senior professor was so old, he would have decided against it." X-ray astronomy was too new in those days and too far outside the mainstream to receive wide support.

Tradition and traditional fields of research play a role in determining the course of science in America as well. Here as elsewhere there is establishment science and science outside the establishment. Perhaps it is no accident that Giacconi and his coworkers were all *physicists* rather than astronomers. They had come to astronomy from outside the field. So they understood things the astronomers did not. They understood new techniques: new types of detectors and new ways of doing things. Happily, there were also some things they did not know. The astronomers, for instance, "knew" that the search for X rays from space was going to be expensive and largely futile. They "knew" that most celestial objects emitted no X radiation. Perhaps only outsiders would have disregarded this general attitude and plunged ahead regardless.

It is also striking that Giacconi and his group were not on the faculty of any university nor employed by any government laboratory, the two traditional sources of support for pure science. They worked for a *private business*, a firm dedicated to the fine art of making money in new and ingenious ways. The firm is American Science and Engineering, situated in Cambridge, Massachusetts, just down the street from Harvard Square. It appears that the directors (one of whom was Giacconi) had decided that X-ray astronomy might be made to yield a profit, and if a little science came along in the process, so much the better. Or at least this is one view of the situation. But another view is also possible. Perhaps Giacconi and his colleagues wanted to do X-ray astronomy, and they did not intend to be stopped just because they were not employed in the usual way.

By now it is all very different. By its spectacular successes X-ray astronomy has become part of establishment science. Ultimately, Giacconi and his group left American Science and Engineering and moved up the street to Harvard University, a properly discreet and mainstream institution. They were not outsiders anymore.

*   *   *

Very early on it was decided that the new satellite would be launched eastward from somewhere on the equator. There were a variety of reasons. In the first place, at the equator the Earth's rotational velocity was fastest—a good 1,000 miles per hour. A satellite launched eastward could take advantage of this velocity in achieving orbit— it already had a head start. Second, NASA maintained a network of tracking stations strung around the equator. A satellite launched from the equator would stay above it forever, continually passing over these stations. It was going to be a little easier to patch together the stream of data telemetered down if it was always received by the same stations and in the same regular order. Finally and most important of all was the noise contaminating the signal the satellite would be attempting to detect. The jamming of a detector by extraneous inputs is always a problem in astronomy. In the optical region of the spectrum *sunlight* is the noise. So we observe the stars at night. In the radio region Huguenin's noise had come from automobiles and radio and television stations. So he put his telescope out in a reservoir. And in the X-ray region of the spectrum the Van Allen belts, clouds of cosmic radiation encircling the Earth, could jam the detectors causing spurious signals. An orbit passing everywhere over the equator would avoid these belts most easily.

Ultimately the satellite was launched from Kenya. The Italian government operated a launching platform there—the San Marco platform. It was convenient and correctly situated, and faced east into the Indian Ocean. Perhaps the fact that Giacconi was originally from Italy also played a role. Certainly it was a fitting choice.

No one names a baby before it is born, and similarly NASA has evolved the tradition of not finally naming a satellite until it is safely launched. After all, the rocket might blow up on the pad. Or the instrument might achieve orbit but fail to operate correctly, in a kind of cosmic stillbirth. Before a satellite is known to be safely up and in good working order it is given only a code name: in this case SAS-1, for Small Astronomy Satellite number one. But as the time drew near, it was decided to launch SAS-1 on the seventh anniversary of Kenya's independence. And a name was chosen: "Uhuru," the Swahili word for "freedom."

X rays, like light and radio signals, are waves in the electromagnetic field. The difference is wavelength. Among these three, radio waves have the longest, light an intermediate, and X radiation the shortest wavelength. Each is detected by different means. Light waves are captured by the rods and cones in the retina of the eye or by photographic film. Radio waves are received by an antenna. As for X rays, dentists and doctors detect them with photographic film. But this is a crude technique. The American Science and Engineering people knew a better one: the proportional counter.

Their proportional counter would be a box containing some gas and a wire. If an X ray passed through, it would briefly ionize the gas. The ions would collect on the wire and produce a pulse of current. They would detect this current: it would report that an X ray had passed through. Geiger counters work the same way.

The pulse of current could not tell which direction the X ray came from. A box full of gas would respond to X rays coming from any direction. How to point the counter? They did this, first, by shielding five out of the six sides of the box, so that X rays could only enter from the front. Next they mounted a collimator on the front: a metal shield that blocked out all X rays except those coming from a particular direction. It could have been as simple a thing as a length of piping. Just as one peers through a pipe, so the proportional counter would "look" through its collimator.

As it slowly took shape in the laboratories of American Science and Engineering, SAS-1 was a bit smaller than a person: a few feet wide and four or five feet high. The heart of the instrument was the proportional counter, roughly the size of a book. Its front wall, through which the X rays would enter, was an ultrathin sheet of beryllium, transparent to X rays. The collimator was a rectangular array of metallic slats. When ultimately assembled the proportional counter looked for all the world like an automobile radiator grille. There were two of them, mounted back to back on opposite sides of the satellite. Uhuru was going to look in two precisely opposite directions at once.

The satellite would slowly spin. Once every twelve minutes it would swing about in its steady rotation. In these twelve minutes each proportional counter would sweep out a narrow band in the sky. Over and over again they would cover the band until, upon command from the ground, the satellite would swing its axis of rotation about, enabling them to scan a different band. A magnetic sensor on board would monitor the Earth's magnetic field, providing information on the direction of the spin axis relative to magnetic north. The position of the satellite in its swing would be monitored by a phototube peering outward through an N-shaped slot. As each star swung by it would give a triple pulse of light as it passed across the N; a constellation a complicated pattern of blips. The blips would be telemetered to Earth. Down below a computer would have stored within its memory banks a detailed map of the sky. By consulting this map it could decide from the pattern of the blips just where the satellite was pointing. Throughout the age of exploration mariners had navigated by compass and by sextant: now, in the new age of the exploration of space, Uhuru would too.

The satellite would draw its power from an array of solar panels. Folded in during launch, they would snap out into position when a safe orbit had been achieved. The entire experiment—the proportional counters, telemetry operation, housekeeping functions, everything—would operate on a mere 30 watts, less than the power expended by a light bulb.

When finally assembled the instrument would be wrapped in an ultrathin layer of aluminized mylar foil. The foil, transparent to X rays, would protect it from wild fluctuations in temperature as it passed from the shadow of the Earth into the heat of the Sun in space. Only the star trackers, a Sun sensor, and the solar panels would project through.

Nine hundred miles high, and traveling endlessly at 17,000 miles per hour . . . but there would be no sensation of speed. Not a breath of wind, not the faintest sound, not a trace of vibration would disturb SAS-1 as it automatically performed its functions. To one side—above, below, to the left, to the right: it makes no difference in space—would be the bulk of the Earth: a panorama, a billboard, curved at the edges, brilliant blue and flecked with white. Steadily the clouds, the lakes, the continents, would drift past. And the satellite: wrapped in foil like some oversized chicken in a roaster, spinning just barely perceptibly, solar panels deployed like a cosmic windmill. A celestial Don Quixote would have made a run at it.

Designing a satellite is a science. Actually building one is an art.

Tananbaum: "The windows on our proportional counters were all made of beryllium and they were fantastically thin. The very first time I picked one up I managed to put a screwdriver through its window. I remember thinking as it broke that it was just like an eggshell. We had to be very careful in dealing with those things. The least amount of water on them, from humidity, for instance, or from your fingertips, and the beryllium would become oxidized. It would rust, in effect. The rust would leave a tiny speck of porosity, a very small pinhole in the window, and all the gas inside would then be contaminated. These counters were very sensitive and we wouldn't get useful signals out of them if there were even minute leaks of this sort. We tried first to coat the beryllium with all sorts of protective materials: resins, epoxies, of different sorts. Then we would put the counter in a tank and heat it up to 100 degrees Fahrenheit and 100 percent humidity. Those were the conditions down at the launch site, pretty nearly, and we needed to know if the instruments could survive them long enough to get into orbit. The tests were done several years before launch and we learned that after even as little as a week in that kind of environment the counters would become completely contaminated. We eventually concluded that we could not come up with a program to reliably coat and protect the counter windows. So from then on we had to wear gloves when handling them, and down at the launch site we were forced to keep the entire satellite permanently sealed off from the environment."

They had problems with electrical discharges within the electronics. "A high voltage on an electronic component in air will basically be fine. That high voltage in a vacuum will also be fine. But there is an intermediate region at which you get strong discharges. Different NASA centers had different approaches to this—some even had different approaches at different times. We flew everything potted. Potting a high-voltage component is simply an attempt to force all the air away from it before you fly by filling it with an epoxy. You put the component in a vacuum chamber and evacuate it and then you mix your epoxy with the catalyst that causes it eventually to harden and you pour it over the component. If you have designed things right everything gets coated and 100 percent filled, and then all of your high voltage is either in cables or running in between regions that have this potting material at the ends and there is no possibility of a discharge.

"The problem is there is something of a black art in actually doing this. You need a very skilled practitioner, and even the best can make mistakes. Perhaps the material wasn't primed quite right, or perhaps it was applied wrong. Then, instead of getting a perfect adhesion of the epoxy to the electronic component little spaces are left, little pockets of trapped air. So then you put the instrument into a vacuum for two days or seven days and the pocket of trapped air gets thinner and thinner and suddenly you hit this critical region and electrical discharge sets in. You get a tremendous amount of noise and destruction of the electronics. Had this occurred in space it could have entirely ruined the satellite. So we had to pot, and then test in a vacuum for a week, and then take out the component that failed and isolate the failure and report it and put it all together and test again.

"There was a guy down at the launch site specifically sent from NASA. He was a specialist purely and simply in cleaning the instrument. He had the job of sitting there the night before it was finally encapsulated and with special lights picking up every last speck of dirt and all the fingerprints and removing them. He was very much a specialist.'

Specialists, engineers, master technicians . . . they labored over SAS-1 endlessly. It was becoming a triumph of craftsmanship—a jewel. But there is a closer analogy. More than anything else SAS-1 was a modern counterpart of the great religious engineering efforts of the past such as the Gothic cathedrals or the pyramids of Egypt. Like them, it required financial resources far beyond those available to an individual: SAS-1 was paid for by the modern secular state; the pyramids by the theocratic state of the Pharaohs; and the cathedrals by the Catholic Church, itself a form of government in those days. Like them, it was a group effort extending over long periods of time, involving large numbers of highly trained technicians backed up by still larger numbers of unskilled laborers.

And finally, it was utterly useless from any strictly practical point of view. Its justification was not found in the realm of the practical. It was an expression of a need—a yearning. The Egyptians built pyramids, the Europeans cathedrals. We do science.

Differences come to mind. The Gothic cathedrals were meant to be seen. They remained in the villages that built them, there to inspire religious awe in generations to come. So too with the pyramids, which have survived for millennia. SAS-1, however, was built to be thrown away. Once declared perfect, it would never be seen again—it would be launched, briefly used, and then burned up upon re-entry when its orbit degraded several years later. A strange intent for such a work of art!

But is it really true that religious art was meant to be seen? We believe so now as we view it in the museum, but that is not its intended place. The pyramids were not built to be appreciated. They were built to conceal. Their function was to protect the dead Pharaoh and his treasures from tomb robbers. As for the treasures themselves, they were carefully fashioned only to be sealed away, if was hoped, forever. And as for the Gothic cathedral, its large-scale proportion and the general plan of the building are impressive enough, and a multitude of sculptures, inscriptions, and stained glass windows take one's breath away. But there is more to it than meets the eye. Elaborately carved gargoyles have been lifted hundreds of feet into the air, perched on dizzyingly high spires, and secreted away everywhere in nooks and crannies. They are too high to be appreciated—they can barely be seen. The same is true of stained glass windows, some of which are conveniently low down, but others, equally fine, far out of sight.

It is almost as if these works of craftsmanship and art were never meant to be seen. Perhaps they were only meant to exist. It is a style of construction that would horrify modern architects with its wastefulness. But it would not horrify the people who built SAS-1.

SAS-1 was launched from the San Marco platform on December 12, 1970. The platform sits in the Indian Ocean three miles off the coast of Kenya, not far from Malindi. Three miles is no accident: it puts the platform in international waters just outside the territorial jurisdiction of Kenya—just in case. "It was a Texas tower sort of thing," Tananbaum told me. "It looked like an oil rig. For all I know, that is what it actually was. There were two of them; the San Marco platform, from which the rocket was launched, and a second which contained the block house, a cement bunker in which the people who controlled the launch were located. The platforms were not that big, each the size of a football field, let's say, and they were a few stories up above the ocean. You went out to them by ferry from the mainland, and you would step out of the boat into this rope cage with a wooden floor and there was a hoist which would lift it up onto the platform. Or sometimes you would go out on a little rubber raft with an outboard motor. You would get a good bumping that way, and they couldn't lower the cage then because there was no place on the raft they could set it down. So you would have to climb up sort of a rope ladder, an interweaving rope net of the kind you might find on a child's jungle gym.

"Once you were out there there was no sensation of being at sea. It felt like you were in a building. The platform was very stable and there were a number of rooms. It was hot, of course, and humid. Everybody went around in shorts and sneakers and without a shirt."

The satellite arrived on the platform one month before launch. Three people from American Science and Engineering went with it, Harvey Tananbaum as the scientist, an electrical engineer, and a mechanical engineer. Also there were people from Johns Hopkins who had built the support vehicle—the power system, the communication system, the electrical system—people from the NASA Goddard Space Flight Center from which the program was managed, various contractors, and the Italian crew that actually did the launching of the rocket. "One month was fairly detailed in terms of the tests we had to do." Tananbaum told me. "The experiment had been left unattended to for a period of time while it was being shipped down, so we set aside about a week in which to unpack it, set it up on a table, set up computers, and check it out. We had to do a certain number of functional tests to make sure the shakes and rattles of the trip hadn't resulted in any damage.

"We ran tests to look for electrical comptability with the launch rocket. Some of our equipment was going to be running at the time of the launch and we wanted to make sure that it wasn't going to generate some signal that the rocket would interpret as a command to fire its engines prematurely or shut them down halfway up or anything like that. We were communicating with the satellite out of a telemetry van on the shore and we went through a whole series of dry runs of the final checks that would be made in the final few hours before launch. We wanted to make sure they could be run in the time that was set aside for them and that we could read the data from the satellite and understand what it meant.

"Sometimes everything would be going so smoothly that we had two or three days free and we would go out to visit the game preserves. Other days we would be through by ten or eleven o'clock in the morning and we would start drinking early. But in spite of all the free time it was hectic. It was pressure-loaded.

"About ten days before launch the people from Johns Hopkins decided there was anomalous behavior in one of their batteries. The main battery didn't seem to be holding a charge properly. They had brought along a replacement and some tough decisions had to be made as to whether to make the switch. The switch was finally made, and it took about a day to do it. Then we had to repeat a number of electrical tests to make sure that everything was working properly. It later turned out we had made a mistake there. Those batteries were very weakly magnetized and for some reason the magnetic field of the replacement battery differed from that of the old. There had been extensive magnetic testing of that first battery, and all of its field had been canceled out by strategically placing small magnets about it—like balancing a tire. When we switched batteries nobody thought to rebalance them. We just assumed they were interchangeable whereas in fact they were not. Once the satellite was up, that faint unbalanced magnetic field interacted with the Earth's magnetic field and produced an uneven rotation of the satellite. It made it spin slightly faster and slower as it went around."

At first the satellite sat in an air-conditioned room to protect its proportional counters from the corrosive effects of Kenya's high humidity. Now, about a week before launch, it was fitted into the nose of the rocket that was to lift it. From this point on, it could no longer be air-conditioned. They kept it dry by pumping nitrogen gas from a tank through a bag surrounding the experiment.

By this point Giacconi had arrived. "It wasn't absolutely necessary that I be there," Giacconi told me, "since everything was going exceedingly well. And there were lots of things going on back home. At this very time we had just completed a major proposal to NASA for a second and far bigger satellite. Actually I debated with myself whether I would stay at the Goddard Space Flight Center where the data would be received. But in the end I just couldn't stand it anymore and down I went.

"Once I got there it turned out I was able to do a few useful things. The fact that I was originally Italian helped since the launch crew was Italian. I could use the bilingual approach, you see. For instance, the crew had to flush that dry nitrogen over the detectors 24 hours a day. It was a terrible nuisance because it meant that somebody had to go out there in the middle of the night to change the tank. And they did it. They did it out of . . . well, they were happy there was this Italian guy they could talk to. A couple of bottles of wine, you know—things like that."

Two days before launch the fuel was loaded into the rocket. From that point on it was armed—a live bomb—and access to the San Marco platform was restricted. "There were several inadvertent things that could have happened," Tananbaum told me. "The thing could have exploded. Only people who had been through a special safety course were allowed onto the platform and I had not. From that point on all our communication with the experiment was from a distance." They never saw it again. Two days later, in the middle of the night, a technician removed the last of the safety devices whose function was to prevent the accidental firing of the rocket engines. He was the last person to visit the platform. All this time the rocket had been lying on its side, SAS-1 packed away in its nose. At midnight, after the last technician had left, the rocket swung vertically upward to point at the sky. Tananbaum was in the telemetry van on shore. Giacconi was on the control platform. Launch was scheduled for dawn.

They waited.

Tananbaum: "There was a native village about a hundred yards away from the telemetry van. The final experiment tests were run in the last hour. I had communicated with Giacconi by telephone telling him that the final tests were nominal. He asked me a few questions and then I was basically through. I went outside the van to watch the launch. And there I was with this trailer with the computers and the air conditioning and the telephones and the telemetry, and not 100 yards away there was a fence and on the other side of the fence there was a native village. Those people lived in grass houses and they had no plumbing. They cooked by fire. They had no electricity. And I could not begin to conceive of the shock, the cultural shock, they were going to feel when there would be this loud noise and the explosion and the rocket would go up.

"The people there were fishermen, and there is a picture I have of a couple of them coming home to the village carrying some fish. They are walking right in front of some bleachers that had been set up for local VIP's to watch the launch. Evenings I used to sit out there and watch the sunset and the rocket and not be sure of which world I was part of. It was a strange time and a strange place. Much of Kenya is modern, you know. The cities are developed, there are automobiles, there are roads, there is electricity. But there is also this other world. Kenya is a very pleasant place to visit and I may never get a chance to go back to Africa again. It was beautiful and it was a precious experience."

The rocket, vertical now, was illuminated by floodlights. Dawn came and the scheduled moment of launch.

Nothing happened.

Giacconi: "The rocket had been provided by a private contractor and they had some representatives out on the control platform. One of the tests they performed on a gas jet that regulated the aspect of the second-stage engine seemed to have failed. They repeated the test and this next time it came out positive. They repeated that test *nine times* and it was positive. And then they insisted on a signature by NASA officials saying the test had been all right. The point was this company had a no-fault contract to provide the rocket for NASA, and it would have given them $100,000 more profit if it had performed without a fault. The slightest problem and they would have lost that bonus money. So now they wanted their signature, and not just from anybody. It had to be from a NASA contracting officer.

"The problem was that there was no contracting officer out on the platform. They were all asleep in the hotel in Malindi. So these guys went off to get them in the middle of the night—three miles by boat on the sea, and then ten miles by bumpy dirt road. And the whole countdown had to stop while they did this."

Giacconi just lay down on the hard steel plates of the platform and went to sleep. "There was nothing I could do so I didn't do anything," he said. "I was simply waiting. As a matter of fact, I didn't really know what was going on at the time. It was all very confidential. I was very tired and I was wet from the sea, the humidity. One of the Italian crew gave me his shirt and I actually managed to sleep. The only thing I wanted to do was get my instrument away from all these meddling fools."

The Sun came up. The satellite, perched in the nose of the rocket, began to heat up. The humidity rose. Deprived of its air-conditioned environment, deprived of its flow of dry nitrogen gas, tiny droplets of water began to collect on the fragile beryllium windows of the proportional counters, setting in motion the irreversible process of decay that soon would ruin them. "The more urgent problem, though, was that we were running out of liquid oxygen," Giacconi told me. "It was necessary for some of the various functions. Now you don't just go down to Malindi and buy some liquid oxygen. If we didn't launch soon we were going to have to abort and this would have meant canceling the launch for maybe two months while the stuff was sent down from Italy. It was touch and go until those guys finally came along with their signature and we could go ahead." Hours had passed.

Tananbaum: "In the final stages I think that I alternated every two minutes from inside the van to out. I would go inside to see if the communication with the satellite was good and then I would go outside to take a look. It was really just a form of trying to break the tension, of course. A few minutes before launch I simply went outside and stood there by the side of the trailer. They did the final countdown over loudspeakers: 10 . . . 9 . . . 8 . . . 7 . . . all the way down. The rocket went off with a flash of light. It was very bright. The van was so far away from the platform that it took 15 or 20 seconds before the sound reached us. It started out dully and built up to a roar."

Giacconi was closer. "It is a tremendous noise. It shakes you and you feel it deep inside your body. It starts off like a low rumble and then it becomes higher pitched and more metallic."

Tananbaum: "The rocket starts to lift very slowly. It looks like it is straining to get off the ground at first. And then it begins to move up. It accelerates.

"I *spoke* to it as it went up. I waved to it. It was a very emotional time. We were all exhausted. I had been awake for a day and a half and I had been at that van for maybe 30 hours. A certain part of me was sufficiently choked up that there must have been a tear in my eye."

Giacconi: "There is so much riding on it—so many years. Watching it rise is like an emotion that leaves you feeling drained when it has gone. But it is going and it is—it is some kind of culmination.

"It left a wonderful trail in the sky. At first you can hear it as it goes up and up, but then it grows fainter and fainter. You can see it for longer than you can hear it because the burning of the engine is very bright. For some reason it left a corkscrew trail of vapor behind it as it went."

Suddenly the rocket engine stopped.

Tananbaum: "I could still see it from the ground. It looked like an airplane seen from a distance. My heart jumped right into my mouth. Then the rocket started again and only then did I remember that it was a multistage engine and this was simply the first stage shutting down. It dropped off and the second stage ignited and it went 'poof' and it took off again. It grew fainter and fainter. Just as a tiny spot I could actually see the second stage shut down and the third stage ignite. And that was as far as I could see it. I went back inside the van."

It was all over in a matter of minutes. The San Marco platform was empty. Four years of planning, three years of building, a month of final testing here in Kenya. But now SAS-1 had vanished. It had been thrown away. It had turned into Uhuru.

"I remember thinking as it went," Tananbaum said to me, "that we had put so much work into it. And now it was time for it to work for us."

Giacconi: "The minute it went the Italian crew became very happy and started breaking out the champagne because the launch had all gone so well. In fact it achieved a very beautiful orbit. But I was nervous as hell because I didn't know if the satellite was working. It didn't mean anything to launch it, not a goddamned thing. The experiment had to *work*. So I was very nervous and I telephoned mission control headquarters in Maryland to see if we couldn't turn the satellite on just briefly as a test. They told me they wouldn't turn it on for a couple of days because they wanted to check if some other things were working, they wanted to deploy the solar panels, and things like that. In other words, they were telling me not to bother them because they were busy. I just couldn't stand it. Luckily, Marjorie Townsend, our program manager from NASA, was with me, and she couldn't stand it either. We decided to cheat a little and turn it on ourselves from the telemetry van.

"There was a rubber raft with an outboard motor moored at the base of the platform and we took off. The guide drove us over to the van. It was three miles away. And while we were sitting there bouncing along and getting wet, the satellite was coming around. It was coming all the way around the world and it was a horse race, it was a real question to see whether we would make it to the van before the satellite passed overhead.

"We made it in time and with Marjorie's consent we turned the instrument on. I looked at it and I could tell that it was working. It was counting X rays. Then we turned it off."

Giacconi leaned back in his chair in his office at Harvard and gave me an enormous grin. "Just kids, that's all. We were just kids. But I knew then the satellite was working. I was happy."

The Binary X-Ray Sources

Two weeks later, Uhuru failed.

The engineers at NASA decided it was a simple problem of overheating. Perhaps the satellite had baked too long on the pad at Malindi, waiting for the launch. Right now it was oriented broadside to the Sun. They sent up a series of commands causing it to swivel sideways, till as small an area as possible was presented to the Sun's rays. The instrument cooled. Within hours they were back on the air.

One month later it failed again. Uhuru accumulated data from its observations by recording them upon a tape. As it passed overhead, a tracking station in Quito, Ecuador, would command it to rewind the tape and play it back. But this time the recorder did not respond.

They repeated the command a number of times and tried a series of other options. No response. The tape recorder was permanently jammed. Data from most of each orbit was now irretrievably lost. Still available, though, was that actually gathered during the time Uhuru was in contact with Quito—about eight minutes out of each ninety-minute orbit. Not so much; on the other hand, eight minutes was the length of a good rocket flight. They were getting the equivalent of a rocket experiment every hour and a half.

Over the months NASA brought in a number of other tracking stations spread around the equator: one in Singapore, another in the Seychelles in the Indian Ocean, Malindi, Ascension Island in the Atlantic, French Guiana on the coast of South America. Each station recorded the data as Uhuru passed over, then mailed the tape to the

Goddard Space Flight Center in Maryland. There engineers patched the bits and pieces together and sent it up to American Science and Engineering. Now they were getting half of each orbit.

The transmitter grew balkier—overheating again. They babied it along, orienting the satellite for longer and longer periods of time sideways to the Sun. It was severely limiting the portions of the sky accessible to them. Things worsened: ultimately only Quito, the best station, was capable of receiving the signals and then barely. They were down to two or three minutes per orbit.

And then, completely by accident, someone at NASA resurrected the transmitter. It was capable of sending at two power levels, high and low. Another satellite was due to be launched and, purely for engineering purposes, its designers needed to know how long it could remain pointed at magnetic north. NASA decided to experiment on Uhuru. A series of commands was set up causing it to swivel around to the north. Inadvertently, the commands also had the effect of switching the transmitter's mode. For some reason it jumped back into operation. They were on the air again.

Meanwhile the X rays were pouring in.

Figure 54 shows an Uhuru scan across Cen X-3, the third X-ray source discovered in the southern hemisphere constellation of the Centaur. The smooth "triangular" rise and fall of the signal was caused by the rotation of the satellite carrying the source across the collimator's field of view. It was the rapid bursts superposed on the trend that belonged to the source itself. They were regular. Cen X-3 was an X-ray pulsar.

The pulses arrived once every 4.8 seconds—a little slow compared

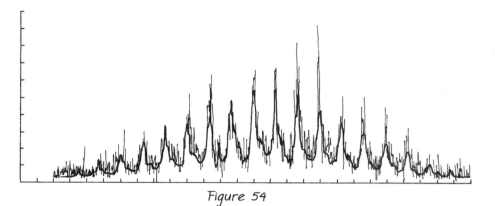

Figure 54

to the radio pulsars, though not unduly so. But here the similarity ended. Unlike them, Cen X-3 emitted no radio pulsations at all. It was not slowing down but speeding up. And it regularly oscillated between an "on" and "off" state. For nearly two days the pulsations persisted; then, quite abruptly, they faded to invisibility. The source remained off for half a day and then, equally abruptly, the pulsations resumed. The on-off pattern itself was highly regular, the cycle repeating exactly once every 2.0871 days. The X-ray pulsar was more complicated than a radio pulsar. It contained not one but two clocks.

The ticking of the first clock was modulated by that of the second. The X-ray pulsations were not quite regular. Over half the two-day cycle Cen X-3 burst a little more rapidly than once per 4.8 seconds; over the next half more slowly. The rate of pulsation smoothly varied—with exactly the same period as the on-off cycle.

Why did a neutron star which emitted X rays behave so differently from those that did not? This regular modulation was the clue. The minute the American Science and Engineering group found it they recognized its significance. It came from the Doppler effect. Years before Hewish had searched for this very phenomenon in the first radio pulsar. He had been looking for the effects of motion—a possible orbital motion of the pulsar about a star. He did not find it. Now, years later, Uhuru had discovered the effect in a completely different source.

Cen X-3 was a neutron star in orbit about a second star. It spun once per 4.8 seconds and emitted a beacon of X radiation. It orbited once every 2.0871 days and, alternately approaching and receding from the Earth, its apparent pulse rate was modulated by the Doppler effect. The abrupt transitions from "on" to "off" states were eclipses: the pulsar passed behind the second star in its orbit. A binary system.

The binary nature of the Cen X-3 was the key that led Giacconi and his colleagues to an understanding of its strange anomalies. A "year" two days long implied a tight orbit—tighter far than Mercury's, which required 88 days to circle the Sun. An eclipse lasting half a day out of this two-day cycle implied that the star which obscured the pulsar—the companion—must be fairly large. Were it a white dwarf, neutron star, or black hole the eclipses would have occupied a far smaller fraction of the orbit. Cen X-3 was a neutron star in orbit about a normal star.

In close orbit—and it was very massive. When one star was so near another it could have a significant effect upon its structure. It could raise a tide.

The Moon lifted a tide upon the Earth. As it passed overhead the ocean surface rose toward it, pulled upward by its gravity. The rise was not that much—a matter of feet—and after the Moon had swung by, the ocean dropped back down again. But this was because gravity from the Moon was weak. Two things might contribute to making tides on other bodies larger: the orbiting object could be more massive, or it could be closer in.

This combination was fulfilled in the binary system containing Cen X-3. On the scale of such things the neutron star was very close to its companion. The stars comprising ordinary binaries were usually separated by billions of miles: the Cen X-3 system was only a few million miles in extent, nearly a thousand times smaller than average. So the tide raised on the companion's surface by the neutron star was correspondingly larger. It did not subside after the pulsar swung by. It lifted completely off the star and into space.

The picture that emerged from all this was of a companion whose shape was radically distorted by gravity from the neutron star. It was not even remotely spherical. The side toward the pulsar puckered outward. It rose in a cusp, and from this cusp streamed a white-hot plume. Welling up from the deep interior of the star, drifting sideways across the surface, currents smoothly carried matter into it. As the gases approached the plume's base they grew lighter. Gravity from the star below was being compensated by that from the pulsar above. They lifted, gently broke free, and poured outward into space. They were falling up.

Up and onto the neutron star above. The pulsar was gaining while the companion lost. The process was known as accretion. If the pulsar were not orbiting, the companion accretion would have been simple, but because it was, things were more complex. The stellar material rose toward it but it was drifting sideways. It sidled out of the way. The accreting material missed the mark, shot past the pulsar, then swung about it in an arc. It fed into a gigantic disc spiraling about the neutron star: the accretion disc. An artist's conception of the system is shown in Figure X (photo section).

Within the disc each parcel of gas was in orbit about the neutron star. But the orbit could not have persisted for long. Turbulent motions—a churning within the disc—would carry material inward. Viscosity—the rubbing of adjacent layers—would damp out shearing motions and do the same. The accreting material was smoothly spiraling inward toward the neutron star.

Ultimately it would have come in contact with the pulsar magnetic field. Like the pulsar magnetosphere, the accretion disc was an ionized plasma, and it was not free to move across the magnetic lines of force. It was guided to move along them. Close to the star it weighed a good deal but it could not fall straight down upon it.

Indeed, at the magnetic equator it could not fall at all. Rather it funneled sideways toward the two magnetic poles. Only there were the lines of force vertical. Only there was the accreting matter free to fall.

And fall it did: helplessly plunging at nearly the velocity of light under the powerful pulsar gravity, smoothly streaming downward along the lines of force—two perfect tubes of violently compressed, superheated gas. Because they were hot they radiated: because they were very hot they radiated not light but X rays.

Two "hot spots" of intense X-ray emission on the neutron star. Two beacons. The star spun, the beacons whirled about. An X-ray pulsar.

Cen X-3 is one of a large class of neutron stars in binary systems discovered by Uhuru. None shows radio pulsations. Apparently the generation of radio beacons in pulsar magnetospheres is a fairly delicate process and in the binaries it is crushed—literally—beneath the weight of the accreting material.

Nor is it surprising that the binary pulsars are speeding up. The acceleration disc from which infalling matter spirals down rotates in the same direction as the pulsar spins. When the matter lands it gives the neutron star a sideways blow, increasing its rate of spin.

What is surprising, though, is the very existence of these binary systems. By rights they ought not to be there at all. The supernova explosion which forms a neutron star is so powerful it stands a good chance of entirely destroying any companion. After all, the systems are unusually tight—the companion lies uncomfortably close to the site of the blast. At the very least the hammer blow of the explosion should break the gravitational bond holding the binary system together, detach the pulsar from its orbit, and send it flying into space.

Among the ordinary stars from which neutron stars are born, about half are binary. But prior to Uhuru *none* of the known pulsars were members of such systems. This striking asymmetry was evident from the very beginning, and it was taken all along as evidence for the correctness of these ideas. Few believed a binary pulsar would ever be found.

Uhuru forced a change in this point of view. In an elaborate process of backing and filling, detailed scenarios are now constructed, complex evolutionary sequences, which allow such systems to exist. Hindsight is a wonderful thing: they seem reasonable—they work. But no one would have thought of them without the binary X-ray sources prodding them along.

* * *

Nevertheless, once these systems do exist they are useful. They allow a measurement to be made that otherwise would have been impossible. The binary X-ray pulsars are the only means we have of weighing neutron stars.

Return to Uhuru orbiting the Earth. It is orbiting in order to stay up there. Uhuru was given an orbital velocity to prevent it from falling down—to make it fall sideways in a circle. But suppose the mass of the Earth were greater. Then its gravitational pull on the satellite would be greater. The satellite's path would curve downward more sharply, and in order to remain in a good orbit its sideways velocity would have to be increased. The moral is that the magnitude of a satellite's orbital velocity is determined by the mass of the Earth. Conversely, it can be used to *measure* this mass.

The same is true of any binary system. Measure the orbits of the two members and you have measured their masses. But when this is done with the binary X-ray pulsars a surprising result emerges. All the neutron stars have the same mass. They are all 1.4 times the mass of the Sun.

That is a striking number. It is the Chandrasekhar limit. But what does the Chandrasekhar limit have to do with neutron stars? Ostensibly it has no relevance for the pulsars whatever: they are theoretically capable of existing with masses either greater or less than 1.4 solar masses. Nothing would prevent us from building one of half the Sun's mass, or two. But nature never does it.

Most people believe this remarkable fact is telling us something about the pulsar formation process. But what? Is there some property of the supernova collapse of a star that is intimately related to the Chandrasekhar limit? Or does the pre-supernova evolution of a star depend upon it? Is it possible that only evolutionary paths leading to just this mass fail to disrupt the binary system? People would dearly love to measure the masses of isolated pulsars in order to see, but so far this has proved impossible. The binary pulsars have thrown an entirely new light on Chandrasekhar's discovery, and one that no one had ever foreseen. But what it all means is anybody's guess at present.

Another of the X-ray binaries is Cyg X-1, the first X-ray source found in the constellation of Cygnus. In fact Cyg X-1 was one of the very first X-ray sources ever found: it had been discovered in an early rocket flight and monitored thereafter by numerous others. All along there were discrepant reports. One experiment would observe it to be very weak, the next as more intense. On the other hand, each flight used different instrumentation and it was difficult to compare them. It was hard to be sure what was going on.

Uhuru resolved the issue. Figure 55 is an Uhuru observation of Cyg X-1 (the triangular response of the collimator has been subtracted out). The source was variable indeed. It flickered and flared incessantly, the fastest bursts being less than a tenth of a second in duration. It looked like a pulsar.

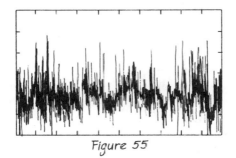

Figure 55

But a more careful look at the data showed some differences. In Figure 55 a slower rise and fall is also evident. The American Science and Engineering group studied this long-term variability by smoothing over the more rapid bursts, averaging the intensity over long stretches of time. Figure 56 shows a series of five-second "time exposures" they obtained in this way. The source fluctuated from one to the next. They tried longer averages. In Figure 57 the exposures are 14 seconds long. Cyg X-1 varied on these time scales as well. It fluctuated on every time scale they looked at.

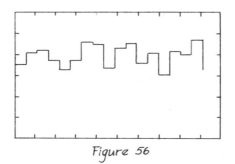

Figure 56

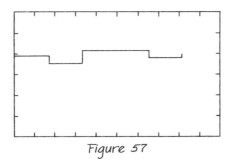

Figure 57

In this regard the object was different from the pulsars, whose beacons, though erratic, possessed an inflexible long-term average shape. Cyg X-1 had no average pulse shape at all. It was absolutely erratic. This turned out to be true of its rate of pulsation as well. The group spent months trying to find an underlying pulse rate in the data but it proved impossible to pin down. They would find a value that fit one stretch of data only to find that it did not fit the next. Cyg X-1 appeared to be constantly changing its rate of pulsation.

Ultimately they abandoned the attempt, and concluded there was no underlying regularity at all. There were only *pulse trains*: brief sequences of regular bursts that would last a few seconds, then subside to be replaced by a train at a different rate. Accompanying this was a continual and erratic flaring with no discernible regularity.

Nevertheless there was a crude similarity with Cen X-3. By analogy, the question whether Cyg X-1 too was binary came to mind. But there was no way to tell. Because there was no basic pulse rate they had no way of looking for a modulation in this rate produced by the Doppler effect. Nor did the source exhibit eclipses. It was always on. This might have meant the object was isolated—but not necessarily. It could be ascribed to simple bad luck. The orientation of the orbit shown in Figure 58 would yield eclipses; that of Figure 59 would not.

Giacconi and his colleagues had reached an impasse. There was nothing more they could do. They published their results and other groups took up the task. Some of these were optical astronomers, but they were hampered by the fact that the Uhuru collimators did not point very well. They did not precisely specify the location in the sky of the X-ray source. Figure XI (photo section) shows Uhuru's crude determination of its position, indicated as a box. It would have taken years to study every star the box contained in hopes of spotting something strange.

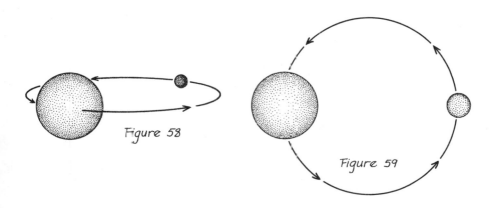

Figure 58

Figure 59

But radio telescopes were far more accurate than Uhuru. If Cyg X-1 emitted radio waves as well, it could be picked out from the maze and its position pinpointed with ease. A group of radio astronomers searched. No luck. The object apparently emitted no radio signals. Then another group searched. They found something. They accurately determined its location. The way seemed open to a good, close look at the object.

Blocking the way, though, was a problem. It was always possible that the radio source and Cyg X-1 were two completely different things. The sky is strewn with radio emitters and it would not have been so unlikely for one, purely by bad luck, to lie close to the X-ray object. But this was laid to rest in a strange and remarkable way.

The first attempt to find the radio source had been on March 22, 1971, the second nine days later. The first group had employed a telescope fully as large and sensitive as the second, and had the source been there they would have found it. The implication was that the source had not been there. The radio emissions must have turned on some time between the two observations. And in the same nine-day span, the X-ray emission as observed by Uhuru underwent a transition. It dimmed to one-quarter its former intensity.

To this day no one knows what happened. Some change in structure must have occurred, causing the object to transfer some of its power from X-ray to radio wavelengths, and the change appears to have been permanent. Luckily, it was not necessary to know. The simple coincidence in time proved the radio and X-ray sources to be one.

At the location of the radio source there was a star. Its name was

HDE 226868. And HDE 226868 was binary. More than that, only one of its components was visible. HDE 226868 was moving in a circle around nothing.

The question was the nature of this "nothing" about which the visible star swung. Cyg X-1 could have turned out to be a completely unremarkable object. Lists of binary systems are actually littered with such cases, and on close inspection they invariably turn out simply to contain one bright and one dim star, the fainter star lost in the glare of the brighter.

A way was needed to test this interpretation. One was available. It relied upon the fact that the mass of the unseen companion (as determined from the orbit) was very large—at least eight times that of the Sun, and possibly greater. But ordinary stars obeyed a so-called mass-luminosity relation: the more massive the star, the brighter it was guaranteed to be. Had the unseen companion been a typical star it would have been easily visible through a telescope. So it was not a typical star. It could only have been a pulsar or a black hole.

Between these two possibilities it was also possible to distinguish— by the very same test. High-mass neutron stars did not even exist. Like white dwarfs, the pulsars were supported by degeneracy pressure and there was a Chandrasekhar limit for them. It worked out to several times the mass of the Sun, far less than the mass of the unseen object.

So Cyg X-1 was not a pulsar either. It was a black hole.

An artist's conception of the binary system containing Cyg X-1 and HDE 226868 would look very much like that of Cen X-3. The black hole orbiting the star draws a tide off its surface. The material streams across to form an accretion disc about the hole and there spirals inward. The deeper in it gets the hotter it becomes, ultimately growing so superheated by friction and compression that it radiates X rays. The crucial difference between a binary system containing a black hole and one containing a neutron star is the absence in the former case of a rapidly rotating magnetic field. The pulsar field imposes order on the accretion disc, and thus order on the emission. In systems such as Cen X-3 it channels the infalling matter into tubes, and produces tightly beamed radiation. What we receive is a uniform series of pulses. But in Cyg X-1 this order is absent and the radiation more chaotic.

The continual and erratic flaring of Cyg X-1 must signal a violent turbulence within the disc, a perpetual shocking and buffeting of the emitting gas by blows from every side. The pulse trains, though, must be something else. They can only arise from local "hot spots," relatively long-lived autonomous structures within the accretion disc. As each moves inward it accelerates, ultimately approaching the velocity of light in its orbit, and the most rapid among them must be coming from near the black hole's edge. In the fastest of the pulse trains we are directly observing matter at the very limit of its existence, fractions of a second from annihilation. In these last instants it is whipping about the hole thousands of times per second in its mad rush. Then it touches the horizon. The pulse train ceases. Unseen, invisible, the spiraling path gives way to a vertical plunge into the singularity.

The discovery of the binary X-ray pulsars must be counted among the most important advances in the astronomy of the seventies. But the unraveling of Cyg X-1's true nature immeasurably surpasses this in its potential impact. Ultimately it will make available to us a natural laboratory in which to study matter under the most extreme conditions, and space-time in its most radical departures from the ordinary. Not least among its implications is the vindication of yet another prediction of general relativity, that of the existence of black holes. Once more, in ways he could not possibly have foreseen, Einstein's thought had reached far forward into the future.

The path of this triumph was long and complex. Among all those involved, Giacconi and his colleagues must be given the major credit. In a way it had happened by accident—if an effort extending over seven years and costing $5,000,000 can be called accidental. But certainly they had not been looking for black holes. They were after something else.

But what they did was crucial. The technique that was ultimately used on Cyg X-1 was nothing new. It had been proposed earlier—in 1966, by the Soviet astronomers Y. B. Zeldovich and O. M. Guseynov. It was they who had first suggested studying binary systems with unseen companions and picking out those of high mass. In their paper they had pointed out seven candidate stars that passed the test. Three years later two American astronomers had gone at it more thoroughly and drawn up a bigger list. These were the best bets.

But HDE 226868 is on neither list. No one had realized its significance. It required the creation of a new science to find a black hole.

God's Play With the World

I first encountered Stephen Hawking at a meeting in Boston. It was the midwinter of 1976 and he was speaking on black holes. The talk he gave that day I will never forget. Its effect on me was electric. On the other hand, it was also mystifying. The discovery he was reporting was so remarkable, so revolutionary and so unexpected, that I had no way of assimilating it. Nothing Hawking said could be fitted into the picture of black holes that I had grown used to.

Few of us in the hall understood Hawking as he spoke. I mean this literally. We did not understand a single word. He had been lifted bodily onto the stage and now sat slumped in a wheelchair—motionless, huddled, a tiny stick of a figure. It was shocking to see how emaciated his illness had left him. He did not seem to be looking at us. He seemed to be staring listlessly at the floor as from his lips came a slow, incomprehensible mumble. It did not sound like speech to me. It sounded like nothing I had ever heard.

Hawking's discovery had come at the end of a maze of calculations, a mathematical journey of daunting complexity, and he had done those calculations in his head. Incapable of holding a pencil, he had nevertheless worked his way through a labyrinth the best of mathematicians would have found intimidating. The terrible disease that ravaged his body made that accomplishment all the more wonderful. Prediction is dangerous, but it seems clear by now that Hawking's discovery will rank as one of the great scientific achievements of our lifetime. Startling as it had been, it was the mere tip of an iceberg. His work has thrust an utterly new picture of black holes before us. It has uncovered a deep and previously unsuspected connection between widely differing branches of physics. And more than that, it may point to a new unifying principle in our understanding of nature. Intimations lurk in the wings of great deeds waiting to be performed, incorporating the radical insight Hawking has won.

But I was not thinking of any of this that day. I was not thinking at all: I was sitting silently, lost in a sense of wonder. Hawking had known that few people would understand him, and so had distributed copies of his talk beforehand. Before me on the page were words of the utmost profundity and power. Throughout the enormous lecture hall people were motionless, following the painful, laborious progress of his speech.

The title of the talk was "Black Holes Are White Hot."

Years after he had first proposed the existence of neutron stars, Fritz Zwicky returned to the subject again, but from a new direction. He asked whether smaller bits of neutronic matter could exist—chunks not miles in diameter but inches, bits the size of golf balls or fleas. Zwicky called them *goblins*. It is hard to know how seriously he took the notion, and few people nowadays pay it much attention, but I believe it should be taken very seriously indeed.

A goblin the size of a football stadium would weigh almost as much as the Earth itself; one the size of a pebble would outweigh a mountain. Goblins just barely big enough to see with the naked eye—dust motes—would weigh a million tons. A person collapsed to the neutronic state would be the size of a bacterium.

Hawking has analogously proposed that there may exist black holes of very low mass. They would be even smaller. A black hole the size of a pebble would be formed not from a mountain, but something the mass of an entire planet. One the size of a bacterium would contain the mass of a large asteroid. And if a mountain were to be crushed into a black hole, it would be no larger than an elementary particle.

Such tiny, massive chunks would have very unfamiliar properties. Elementary particles can only be detected with specialized equipment such as bubble chambers and Cherenkov counters, but an elementary-particle-sized black hole passing through a person would be "detected" in quite a different way—it would punch a tube about the size of a needle all the way through. It would *hurt*. The same quantity of matter in the neutronic state would be utterly deadly:

it would drill an inch-sized tunnel through anything obstructing its path. More massive things would have still more climactic effects: a black hole the size of a bacterium would emit a blast of energy comparable to that from a lightning bolt. So massive are such objects that their momenta would effectively prevent anything from slowing them down. A small black hole could penetrate hundreds of light years of solid rock before finally coming to rest.

Are such things even possible? As mentioned in Chapter 1, grave difficulties would be experienced in producing any small chunks of neutronic matter. They are under enormous pressure and will explode unless contained in some way. In neutron stars gravity does the containing, but goblins would have to be held together in another manner. The only possibility is some intense force of attraction between the elementary particles out of which they are composed that might develop at high densities. At present no one has succeeded in showing that this attraction exists, but so unsure are we of the relevant physics that the situation is wide open.

A spray of goblins could conceivably be formed in the very collapse in which a pulsar is created. Certainly matter is crushed to the appropriate density in such an implosion. Alternatively, one can imagine them chipped away from an already existing neutron star. Giant meteors falling upon the Earth may occasionally blast some of their debris away from the planet and entirely into space: meteors on pulsars might similarly launch goblins on an interstellar journey. The pulsar electric field, which accelerates charged particles to nearly the velocity of light, would do the same to any neutronic pebbles lying about loose. One would not want to be standing nearby if any landed upon the Earth.

With small black holes we are on different territory. On the one hand, there is no problem of maintaining them against their internal pressure: the Schwarzschild surface will contain anything. But producing one is a formidable task. To transform a mountain-sized mass into a small black hole, something would have to compress it until all the myriad elementary particles out of which it is made are crushed into a volume of space ordinarily reserved for one —an "overlapping" even the most adventuresome of physicists finds unnerving.

In guessing how this might be achieved in nature, attention naturally turns to singularities—the states of infinite compression predicted by general relativity. One possibility is the singularity at the center of a white hole. Since white holes, if they exist at all, will spew forth chunks of matter from time to time, it would hardly be compounding a felony to imagine them emitting small black holes as well, or bits of matter so concentrated as to collapse soon after their ejection. On the other hand, such a hypothesis lies at the very edge of speculation. We have no reason to believe white holes would not do such things, but neither is there any reason to believe they would. A far less speculative alternative is the singularity, encountered billions of years in the past, in which the universe itself began: the big bang.

The singularity at the center of a hole occurs at a particular point in space but persists for all time. The singularity which was the big bang occurred at a particular point in time—between 10 and 20 billion years ago—but it included every point in space. Nothing escaped it. Each particle of matter that now exists—the electrons in the Moon, the protons in my body—once was crushed into a state of infinite density and temperature. Emerging from this singularity the matter spewed forth in an endless expansion. The more the universe expanded, the colder it grew, and the more slowly the expansion proceeded. Ultimately, eons later, it condensed into stars, galaxies; the galaxies still drifting steadily apart in a ponderous reenactment of the creation. The big bang is a natural arena in which to look for the formation of small black holes—and large ones and goblins, too, for that matter. Fractions of a second after the singularity, the universe was at the required density. One imagines small fluctuations, tiny subunits at a slightly higher density that fragmented off, reversed their expansion, and collapsed inward.

In quantum mechanics we find still another means by which goblins and small black holes might be formed.

According to the Heisenberg uncertainty principle, not just the position but also the *size* of every object is imprecise to some degree. One can think of this imprecision as arising from a continual and erratic fluctuation. All things—acorns, stones—are perpetually changing in size and shape: expanding outward, shrinking inward.

Ordinarily the fluctuation escapes notice. Usually it is significant only for small objects such as atoms, and is invisibly small for everyday things. But once in a very long while a dramatic shift can occur. An acorn spontaneously contracts—not microscopically, but to an enormous degree. It contracts all the way down. The acorn has quantum-mechanically transmuted itself into a goblin or a small black hole.

The rate of this process is exceedingly minute. So rare are the quantum fluctuations necessary to bring about the gravitational collapse of everyday objects that we are safe in completely ignoring them for ordinary purposes. Almost certainly it has not happened once in the entire history of the Earth as a planet. But it has happened elsewhere, for given enough time and enough space even the rarest of events is bound to occur. And even aside from this, the real point is one of principle. In the last analysis nothing is completely safe from implosion. By its very nature each object in the universe contains hidden within itself a time bomb, the seeds of its own destruction.

Stephen Hawking was born in Oxford in 1942, the eldest of four children, and he grew up in and around London. His father was a biologist—interested, ironically enough, in the study of tropical diseases. Hawking returned to Oxford for undergraduate work in 1959 and then enrolled as a graduate student at Cambridge. Throughout these early years he was a good student, even unusually so, but there was little to indicate his subsequent spectacular career.

It was during his first year at Cambridge that the first symptoms of atypical amyotrophic lateral sclerosis, a degenerative disease of the nervous system, began to manifest themselves. He began to stumble and slur his speech, and experienced a loss of strength. Hawking has no idea how and where he contracted this illness. It appeared shortly after a trip to the Middle East but there is no evidence that he got it there. Whether it was communicated by a virus, whether it was some form of autoimmune reaction, or whether it could even have come as the result of an inoculation from a faulty batch of serum—all remain possibilities. Within a year he walked slowly and with a cane, and his speech was growing difficult to understand. A few years later he was confined to a wheelchair. He grew thinner.

The disease is fatal. Most people die within a few years.

Stephen Hawking did not die. He got married and now, twenty years later, he has three children. He did a Ph.D. thesis on cosmology. As the disease got worse his work got better. Hawking's special forte was the application of the most powerful and abstract methods of modern mathematics to relativity. By the late sixties he had established a series of profound and important theorems relating to the occurrence of singularities in cosmology. Colleagues working in the field found themselves almost as impressed by the brilliance and ingenuity of the methods he had used as by the theorems themselves. Still in his twenties, Hawking was establishing himself as one of the foremost theoretical physicists of his day.

In the early seventies Hawking turned his attention to black holes and went on to prove some of the most beautiful and significant results pertaining to them that we have. In 1973 the notion occurred to him that very small black holes might exist, and he spent a good deal of time considering their properties. Although he did not know it at the time, it was this that set the stage for his most important work.

In September of 1973 Hawking spent ten days in Moscow and paid a visit to the great Russian physicist Zeldovich. There Zeldovich told him of some interesting ideas he had been working on with a colleague concerning the interaction of a black hole with light. Hawking came away from that visit convinced that Zeldovich was right and that the matter was worth looking into. But he was also of the opinion that the Russians were not going about things in the right way. He decided to do it better.

In particular, Hawking wanted to include the effects of quantum mechanics on the interaction process. Normally no one pays any attention to quantum theory when thinking about black holes, for the good and sufficient reason that it is ordinarily confined to the realm of the very small. Its effects on large objects should be negligible. Hawking's desire to include it now came about because of his interest in low-mass black holes, which were sufficiently small that quantum mechanics would have dramatic effects on their properties. It was this combination of the two great cornerstones of the twentieth-century physics, quantum theory and general relativity, that led to his discovery.

He got going in the fall of 1973, and it is a measure of his extraordinary technical abilities that within a mere few months he had an answer in hand. But it did not please him. Indeed the results of his calculations seemed nonsensical. His mathematics was saying that the hole glowed. It radiated light. The black hole was acting as if it were white hot.

Hawking regarded the result with irritation and set about discovering his mistake. The calculation had been a delicate one and there was plenty of room for error. He had cut corners, made a number of approximations, and he now asked whether these had contributed to the impossible conclusion he had reached. Equally serious was the matter of his basic method, for quantum mechanics is utterly unlike relativity. The two theories do not speak the same language, so much so that it had been a serious question how to go about joining them together. Did the fault lie there? He fiddled, trying this, trying that.

Nothing he did solved the problem. The pesky hole persisted in shining brightly. Eventually, Hawking began to entertain the notion that he had stumbled upon something real.

Early in January he mentioned his new result to Dennis Sciama, his old thesis advisor, who was organizing a forthcoming conference.

Hawking recalls that at the time he had not known what to make of it all, but that Sciama had taken the result very seriously indeed. Sciama began spreading the word around.

Hawking's birthday was four days later and the family put on a dinner party for him in celebration. As they sat down to a table loaded with food, they were interrupted by the ringing of the telephone. It was his colleague Roger Penrose, with whom he had done some of his most important earlier work, calling from London. Penrose had heard the rumor and wanted to know the details. Hawking explained; Penrose kept plying him with questions. As the rest of the family sat fidgeting at the table, the conversation dragged on and on. Forty-five minutes passed before Hawking finally returned to the table.

The dinner was ruined.

A rod of metal heated by a blowtorch glows a dull red. Heat it more and it glows bright yellow. In either case the emission of light arises because the metal is hot. Hawking's discovery was that to every black hole there is an associated temperature: just like the rod, the black hole glows.

The temperature depends upon the mass, and a black hole formed from something heavy like a star turns out to be fantastically cold: less than a millionth of a degree above absolute zero. At such very low temperatures the emission of light is completely negligible. Thus Hawking's discovery imposes no significant change on our picture of large black holes: they are indeed black. In particular, the description given in Chapter 7 of a trip to such a hole remains valid.

But smaller holes are hotter. One the mass of a large asteroid, about the size of a bacterium, would be at room temperature. It would emit infrared radiation—invisible to the naked eye, but there nevertheless. A hole formed from a lighter asteroid, far smaller, would glow white hot. And a hole the size of an elementary particle would blaze forth gamma rays with a power of a billion watts.

This radiated light carries energy, which must come from somewhere. In the case of the metal rod the energy came from the blowtorch. As for the hole, it comes from its *mass*. Every black hole in the universe is in the process of steadily converting its mass into energy, and radiating it away in the form of light. Thus, as the emission proceeds, the hole decays. It grows smaller—but as it grows smaller it grows hotter, and the emission more pronounced. Ultimately *all* of the mass has been transformed into energy: the black hole has entirely vanished, and in its place is an expanding sphere of light.

Because the emission from stellar-mass holes is so weak, they do not decay in this manner to any appreciable degree. One formed billions of years ago in the big bang would not have decreased its mass by a fraction of an ounce throughout the entire history of the universe. But smaller holes formed in the big bang would have decayed more rapidly. Very small ones would have disappeared entirely by now, and those initially about the mass of a mountain would have reached the end of their lifetimes at the present epoch. At this very moment they are in the final stages of dissolution, and are evaporating in bursts—soundless flares of pure radiance.

By February of 1974 Hawking believed in his result more than he disbelieved in it, and he gave a talk at Sciama's conference. His talk triggered something of a stir. Most people felt he must have made a mistake somewhere: one participant (who shall remain nameless) declared it all nonsense and rushed from the room, pausing only to sweep up a colleague before sitting down at his desk to write a paper explaining why. This paper was duly published in a scientific journal, having first been sent by the editor to Hawking for his comments. Hawking figured that anyone wanting to blunder in public was free to do so, and he recommended publication.

The emission he had discovered violated everything people knew about black holes—including the prior discoveries he himself had made. Black holes were supposed to suck things in and grow: he had shown that they actually shrank. They were supposed to absorb light: he had shown that they radiated it. Far from being dark, invisible objects, they advertised their presence most spectacularly, the small ones at least. Small wonder that it took people time to accustom themselves to Hawking's new way of thinking.

The crucial element he had added to the theory of black holes was quantum mechanics—in particular, the quantum mechanics of the vacuum. The black hole is a warp in the geometry of empty space, and it is this emptiness that radiates light. It is the vacuum—pure nothingness—that grows hot and shines.

In ordinary physical theory statements such as these are meaningless. They make no sense. It is only quantum theory that gives them meaning, and it does so by reanalyzing what we mean by a vacuum. A vacuum is the absence of matter—but is this even possible? According to quantum mechanics it is not. Just as the uncertainty principle imposes a continual fluctuation on the size and position of every object, so it imposes a corresponding fluctuation on the *number* of objects present.

In particular, this number can never be reduced to zero. If we were to pump air out of a chamber in an attempt to create a vacuum, we would find ourselves incapable of removing every last particle— not because they refused to leave, but because they would be replaced by new particles continually created out of nothingness. These particles are created in pairs, and each pair survives only a very short time before vanishing. Once gone it reappears again, then disappears again; incessantly popping in and out of existence. The process is diagrammed in Figure 60.

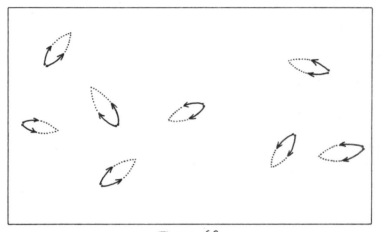

Figure 60

Pairs of all types take part in this fluctuation of the vacuum—in particular, pairs of photons, or particles of light. Nothing can be done to prevent their creation. So perfectly empty space is a figment of the imagination. It cannot exist: the most that can be obtained is this remarkable ocean of sparks.

In ordinary situations the photon pair is so short-lived that this process has no dramatic effects (although it does alter to a measurable degree the structure of atoms). But the black hole is not an ordinary situation. In particular, some pairs will be created in the vicinity of the horizon of the hole, and for these a striking possibility exists. As diagrammed in Figure 61, the pair can be broken, pulled apart by gravity before it recombines and vanishes into nothingness. One member of the pair of photons can cross the horizon, there to be sucked downward into the singularity. This photon turns out to carry a negative quantity of energy. Thus, as it falls in, it reduces the mass of the hole, and when enough have fallen in, the hole has been entirely canceled out.

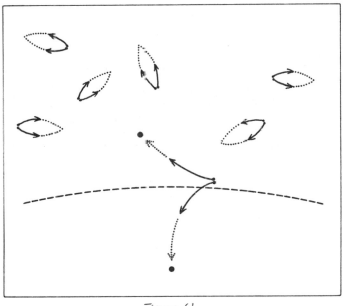

Figure 61

As for the other member of the photon pair, which has remained outside the horizon, it has been deprived of its companion. Now it is no longer able to disappear into nothingness, for such disappearances can only happen in twos: each photon can only reach oblivion by annihilating itself against the other. The bereft photon has suffered a remarkable fate—it has been made permanent. It flies away into space. It is the light "emitted" by the black hole.

After the 1974 conference at which Hawking announced his discovery there was a Great Silence. Before committing themselves to an opinion one way or another the experts went off to repeat for themselves his calculations; trying different methods, looking for an error. Not for about a year did people begin to come around.

During this time Hawking published a brief paper describing in general terms the new radiation, and a longer one discussing his method of calculation. But the 1974 meeting had been so small, so restricted to those few experts actually working in the field, and his first paper so short and the second so technical, that not many scientists paid attention at first. In my own case I can recall being vaguely aware of his discovery at the time, but not until the 1976 meeting in Boston was it forcibly intruded into my mind. My guess is that this was true of other scientists as well.

After 1976 the level of interest grew. Initially it was the fact of the existence of black-hole emission that attracted most attention, and people asked whether the prediction could be tested observationally. The only possibility was the detection of the intense burst of radiation signaling the final evaporation of a low-mass hole. Some groups calculated its properties in detail; others sifted through data searching for examples. Such brief, sharp bursts of gamma rays had, in fact, already been detected; but as things developed they did not exactly have the right properties. To this day nothing has ever been found which can be interpreted as confirming Hawking's prediction.

As for Hawking himself, this does not faze him, for the history of science is littered with too many examples of true predictions unconfirmed for decades. In recent years he has been operating on the assumption that black-hole emission does in fact occur, and he has gone far beyond this question. His present feeling is that his original work may have uncovered a mere tip of an iceberg, and that the real significance of the radiation lies elsewhere. In its broadest terms Hawking's discovery sheds new light on a profound and age-old question: is the universe comprehensible?

Walking by a building against which a stepladder was propped the other day, I happened to witness it fall. Just as I approached, it tilted sideways and gently tumbled downward. Mysterious enough: why did it fall?

Many answers to this question passed idly through my mind as I stepped over the ladder and continued on my way. Perhaps there was a speck of gravel on the sidewalk. Perhaps there was some oil. My footfall may have given just the vibration needed to dislodge it. Alternatively, some tremor in the building against which the ladder leaned may have shaken it loose. All these occurred to me as possible explanations of the event—but it is worth noting that *not once did I imagine that my question had no answer at all.* Though I did not know the explanation, I was convinced that one existed. Unconsciously operating in my mind was the principle of cause and effect: everything that happens does so for a reason.

There are reasons for things because the natural world obeys fixed and immutable laws. The task of physics is to discover them: once found, they enable us to comprehend the universe. This was the program of classical physics as it was conceived of for generations. The nineteenth-century scientist Laplace once wrote that if he were to be given the exact position and velocity of every atom, he would be able to predict—in principle if not in practice—the entire future history of the universe. If yesterday a giant meteor had landed near Chicago, the crater it dug will eventually erode away. But even afterward a detailed accounting of every atom in the world will still reveal traces of its passage. Such-and-such an oxygen molecule in the atmosphere will be *here* rather than *there* because it was shoved aside so long ago. The temperature of the subsurface rocks in Illinois will be microscopically higher than it should have been because it was heated by the blast. A speck of North American dust will be in Chile because it had been exploded upward, launched into the upper atmosphere, wafted about for decades before gently coming to rest in the Andes, frozen into a block of ice, carried downward by a glacier, deposited in a moraine, swept up in a trickle which fed into a stream, and finally deposited beside an irrigation ditch.

It is no matter that in actual practice the passage of an ancient meteor cannot be discovered in this way. Laplace's point was one of principle, and the principle was absolute determinism. The world was predictable: every event had its effects, which triggered still more events in a perfect chain.

The greatest single discovery of twentieth-century physics is that this principle is false. This discovery has two quite different facets, the first of which is the uncertainty principle of quantum mechanics. According to quantum theory, things that happen in the world do so quite without cause. Modern physics can explain in detail why uranium as a substance is naturally unstable, and predict accurately the radiation level of a nuclear reactor. But as to just when a particular nucleus of uranium will decay, quantum theory is silent: the decay could occur this instant, or not for a billion years, and between the two there is no way to decide. Furthermore the decay, when it finally comes, will do so for literally no reason at all. Similarly, an exact accounting of the position and velocity of every particle in the universe cannot be drawn up—not because it is hard, but because there is no such thing. Atoms are not *there,* they are there with a certain probability. Laws of cause and effect are replaced by the laws of chance.

According to the uncertainty principle the meteor's passage did indeed disturb the motion of particles the world over—but because their prior motion was not exactly defined the disturbance had a certain fuzzy quality to it. As the centuries passed, the meteor's effect grew more and more minute, while the uncertainty associated with it did not. Ultimately the uncertainty overwhelmed the effect: the record was blurred and had become lost. Facts had dissolved.

Albert Einstein could never accept the overthrow of determinism in physics and quantum theory's emphasis on the laws of probability.

He summed up his belief in a famous aphorism: "God does not play dice with the universe." By this he did not mean that quantum mechanics was wrong, for it had passed far too many experimental tests for him to entertain this possibility seriously. He meant that it was incomplete; a pale shadow of some deeper, truer theory eventually to be found that would reinstate the laws of cause and effect to their rightful place. But no theory has ever been found, and it is ironic that Einstein, who had taken part in the invention of quantum theory, is also the creator of the twentieth century's second great blow to the principle of causation: relativity, and its child the black hole. The black hole does not dissolve facts. It swallows them.

To demonstrate this I propose to play a game with you. Before the two of us a black hole is floating and I am holding something in my hands. You will turn away as I drop this thing into the hole. Now you turn back again. Your task is to find out what it was I dropped.

Of course you cannot do so. It is hidden by the horizon. Short of jumping into the hole yourself, nothing you can do will uncover the nature of that object—and if you do jump in you will have no way of communicating your knowledge to any third party who remains outside. The horizon of the hole divides space into two regions, an inside and an outside, and things within are absolutely unknowable to an observer who remains without. This in spite of the fact that my secret object is still *there*: intact, unchanged, just inches away from you as it creeps endlessly toward the horizon.

The same is true of the hole itself, which might have been created by the collapse of a star at the end of its nuclear-burning phase— but then again, might not, for it could also have been created in an experiment of crushing some gigantic load of potatoes that went too far. It could even be built out of pure energy, for a beam of light sufficiently intense will collapse in on itself to form a black hole.

There is no analogue for this in the realm of ordinary experience. If I were to drop my object into some sewer hole, or down a cliff, there are ways you have of determining its nature. If an ancient city once stood upon the plains of Kansas, an industrious archaeologist always has the chance of uncovering it and throwing light upon our history. All these things are possible because the ordinary world contains within itself a record of the past: find this record and you have comprehended some bit of the universe. Once within the horizon of the hole, though, history comes to an end. The past is annulled.

*   *   *

A remarkable feature of Hawking's discovery is that it invokes the operation of both these breakdowns of determinism in the natural world. But it goes still farther, for a hole does not simply shine. It acts literally and exactly as if it were *hot*.

Light can be produced in many ways, and depending on the circumstances of its generation it can have many properties. But the light emitted by an object by virtue of its being hot has one feature that renders it of great interest to physicists: *it is in the most random possible state*. Thermodynamics is the branch of physics which deals with heat, and its content is summarized in two great laws. But a far more profound interpretation of these laws is also possible: they are statements about randomness, about confusion. And long before Hawking's discovery there had been intimations that in some mysterious way black holes had something to do with thermodynamics.

These intimations flowed from the work of Jacob Bekenstein, then a graduate student at Princeton University. Bekenstein's Ph.D. thesis showed that there was a remarkably close analogy between the laws of thermodynamics and certain laws relating to black holes. For each statement one could make concerning heat, there was a corresponding statement that could be made about black holes.

Bekenstein's work was an example of the elegant, abstract methods of theoretical physics at their best. It was a beautiful piece of research, but its implications were unclear. Reasoning by analogy is a powerful technique, but often prone to ambiguity. One is free either to accept the analogy or to reject it, and the choice is usually a matter of taste. In this case many workers in the field were impressed by Bekenstein's suggestion, and pushed it further in a variety of ways. But Stephen Hawking had not been impressed. He was of the opinion that Bekenstein's work had no particular significance, and he felt this for a very definite reason. The analogy, he knew, could never be complete. It was approximate at best. And the reason was that hot things emitted light but black holes did not.

It was two years later that he found they did.

Hawking's discovery completed the chain of Bekenstein's argument, and the analogy between black holes and heat became exact. Hawking had not wanted to achieve this end, and Bekenstein had had no way of predicting from what direction the final resolution would come. All in all, it had been a remarkable act of multiple sleepwalking.

Hawking's method of analysis had made no reference to Bekenstein's analogies at all. It was therefore doubly surprising that the radiation he found had just the property of complete randomness that the thermodynamic analogy would predict. But it is not easy to see how this comes about in rereading his original papers: it simply emerges as a result of the calculations, with no intuitively satisfying explanation. Bekenstein's analogy provides an understanding, but Hawking, in his most recent work, is taking the position that a still deeper explanation must be sought. His feeling is that this discovery points to yet a *third* breakdown of the law of determinism in physics.

His reason is the fact that the emission from a hole is as random as possible. This means that we cannot predict exactly what it will be like. We can only predict what it will most probably be like. Ordinarily the hole emits light in a steady glow—but from time to time the emission will fluctuate. A tiny burst of the "wrong" color can occur: a small green flicker now, a big red flare tomorrow. The hole can even emit something other than light on occasion: a stone, a person.

The implications of this can be illustrated by means of an imaginary experiment. Take a large collection of matter and put it in a box. It makes no difference what the matter is: marbles will do as well as anything. Crush them together until they collapse to form a black hole. Now the box contains nothing but the hole.

Stand back and wait. Gradually the box fills up with light, and the hole grows smaller. Eventually the hole disappears entirely: the marbles have been transformed into pure radiance. But not exactly! There may be something else within the box—a ray of light of the wrong color, perhaps, or a tiny bit of wood emitted by the hole. And the problem is that we have no way of predicting what the box will contain once the hole has vanished.

Do it again. Crush the contents of the box, whatever they may be, to form a second hole, and let the hole decay another time. The box eventually fills up with something else—light and a baby whale, perhaps. Repeat the process still a third time, and then a fourth, and so on for as long as we wish. At the end of each cycle the contents are utterly unpredictable. Physics has lost the ability to comprehend the sequence of events within the box.

Hawking is undismayed by this new limitation he has found. On the contrary, he seems almost exhilarated by it. He even has an aphorism of his own, to counter that of Einstein:

"God not only plays dice. He sometimes throws them where they cannot be seen."

Hawking works at Cambridge University in England, in the Department of Applied Mathematics and Theoretical Physics. With the express purpose of finding how a man unable to scribble a note to his wife can do such things, I paid a visit to him there not long ago. The Department sits off a narrow winding lane, and with typical British understatement advertises its presence by a sign so minute that I walked past it fully three times before a kindly stranger recognized my distress and pointed me in the right direction. Passing through an archway, I came upon a cobbled courtyard. Cars were parked about, and a profusion of bicycles leaned against the walls.

The decor inside hardly did justice to one of the most prestigious departments in the world. It was gray, drab, and barren. A linoleum-floored corridor wound this way and that, seeming to pass nothing in particular before reaching a common room onto which a number of offices opened. One was Hawking's.

He sat slumped in a wheelchair at a spacious desk, his arms crossed upon his lap. His head was tilted to one side; occasionally he would lift it up and over to the other side for comfort. During the time I was with him the only other parts of his body I ever saw him move were the tips of his fingers.

Hawking controls his battery-powered wheelchair by means of a small lever upon its arm, and beside his desk are three machines, also operated by fingertip. One is a telephone equipped with a loudspeaker and a microphone into which he can speak without having to grasp the handset. Even those who know him well have difficulty understanding him over the phone, however. I myself found his speech incomprehensible even face to face, and was only able to converse with the help of a "translator"—Ian Moss, a postdoctoral fellow with whom he works.

Hawking's speech takes getting used to. To the untutored ear it sounds like a steady, uninflected drone, and it appears difficult for him to speak at all. This makes it impossible for him to engage in the normal human custom of chatter—of talking loosely of this and that, hopping rapidly from one topic to another. To each of my questions his reply came back very slowly, and punctuated with many pauses. Our conversation seemed almost like an exchange of a series of letters.

Beside the telephone is his second machine—a platform on which books can be placed, and which turns their pages mechanically. Because he cannot handle things Hawking finds reading one of his greatest difficulties. A book the machine can deal with, once someone has set it up correctly: as for a scientific paper, he will have it xeroxed and the pages spread out upon his desk. He spends a good deal of time sitting silently before the printed page, turning things over in his mind.

The third device is a desktop computer. It is specially equipped with two levers designed to replace the ordinary keyboard, and he showed me how it works. At his request Moss lifted him in the wheelchair into a more upright position, then picked up his hands and gently placed them upon the levers. Before him on the screen a representation of the typewriter keyboard appeared, and a small arrow. With tiny motions of his fingertips Hawking pushed one of the levels this way and that, controlling the position of the arrow. When it pointed to the letter "I" he pressed the other lever: the "I" was inserted into the machine. He maneuvered the arrow down to the space bar, inserted it, moved the arrow upward, made an error and inserted a "D," backed up and erased it, inserted a "C" and continued on. The phrase "I can type sentences, and so forth" built up slowly upon the screen.

By the time he finished more than three minutes had passed. He does not use the computer very much.

One of the chapters of Hawking's Ph.D. thesis had been particularly mathematical in content, and was literally crammed with equations. But he told me that it had been the last work of this sort he had ever performed. As the disease progressed he had grown unable to hold a pencil, and is now compelled to work in different ways. He has adopted certain highly geometrical methods of modern mathematics in which equations are replaced by diagrams. "I draw little pictures in my head," he explained. And he spends a good deal of time looking for shortcuts, pretty tricks that will obviate the need for detailed calculation, and those who work with him say that he is masterful at finding the elegant, brilliant solution to problems that others would attack by brute force. Often he works with colleagues, or students—bouncing ideas around, refining his thoughts by explaining them. When faced with a particularly thorny piece of mathematics, the colleague will write it on the board, and the two of them will mull it over together. Often he asks someone to write down for him the subsidiary steps in a calculation.

As Hawking explained all this to me that day, an enormous bafflement possessed me. Nothing he was saying sufficed to explain his accomplishments. Plenty of scientists work in pairs, and use modern mathematics, and look for the brilliant solution. Why was he so different? How had he won through thickets of mathematics, and made himself one of the foremost scientists of our age? I asked if he had a photographic memory. He assured me that he did not. Could he hold in his mind many complex equations? No. Could he multiply 215 by 73 in his head? He could not. Someone has said that Hawking's work can be compared to a Mozart composing entire symphonies in his head. He dismisses the notion with a laugh. When I left him, it was with a feeling of frustration.

But walking along the banks of the river Cam later that afternoon, it struck me that Stephen Hawking had not been withholding anything. It struck me that his remarkable ability was as incomprehensible to him as it was to me. And why should it not be so?—for no one can explain how he thinks. I realized that the operations of Hawking's mind were closed to him, as they are to all of us. Sitting across the desk from me had been a man as enigmatic as the object of his study.

A student zoomed by on a bicycle. I decided that I had come with a foolish question. I had wanted Hawking to tell me How He Did It. I wanted to know his secret. But the only parallel I could think of to his achievement was that of the deaf Beethoven, and the music he had composed. Who was I to take away the wonder of it?

Descent Into the Maelstrom

The Gulf of Corrievreckan runs between the islands of Jura and Scarba off Ireland, not far from the coast of Argyll, to the west of Arran. It is a region of many whirlpools, the biggest known as Cailleach, meaning "old hag." Something happened there many years ago.

A local story tells how Prince Brecan of Norway fell in love with the daughter of the lord of Corrie, a fishing village on Arran. But her father refused to consent to the marriage until Brecan underwent an ordeal. To prove his love, the prince would have to anchor his ship for three days at the site of the whirlpool.

Brecan sent home for three ropes to use in anchoring his ship. The first was hemp. It broke at the outset and had to be abandoned. The second, of silk, was stronger but it too was ultimately torn apart. The last rope was made of maidens' hair—woven from the tresses of one hundred virgins. It was the strongest of them all. But as the tides built up and the whirlpool grew, miles away in Norway one of those maidens lost her virginity. As this very instant the rope parted, and Brecan and his ship were lost.

This legend was recounted to me by Brandon Carter, a mathematical physicist who has specialized in general relativity. He thinks it has something to do with black holes.

In Chapter 11 a "map" of a black hole was drawn representing it as a funnel (Figure 49). The remarkable thing about this whirlpool-like image is the degree to which it has gripped people's imagination—everyone's imagination, scientists and nonscientists alike. Technically speaking, such a map is no more accurate a description than a space-time diagram, or any of a variety of other representations. Furthermore, such diagrams are never able to depict completely the black hole, which is of course invisible. No matter: in spite of everything it has been seized upon to represent in an imaginative way the black hole. It occurs repeatedly both in popular expositions and in the scientific journals. Somehow the picture speaks to us. It is as if the polyconic projection were to find some deep inner resonance in our minds.

Hardly ever has a scientific subject proved so gripping as the black hole. Far more than many other, equally remarkable discoveries, it seems to have struck a responsive chord. Articles and books appear regularly, and at least one full-length movie has come out. Nor is their appeal confined to the general public. Scientists too, in a few short years, seem to have become positively enamored of them. Before the mid-sixties it was difficult to find more than a handful of physicists willing to take black holes seriously. Now the pendulum has advanced full swing and they are invoked incessantly. Each time a new phenomenon is discovered, some lively theoretician is guaranteed to propose that it is caused by black holes in some exotic way. All this despite the fact that after the most intense efforts only one has ever been discovered! In the natural world black holes are hardly ubiquitous. They are only ubiquitous in our minds.

But why? It is not as if they held any particular significance for our present political or economic situation. It is not as if they affected our image of ourselves, as did the Darwinian theory of evolution or recent discoveries in artificial intelligence. What is the human appeal of this ostensibly nonhuman thing?

To me the answer is contained in the imagery we employ. A black hole is a funnel. A funnel is a whirlpool. And in the legend of Corrievreckan, there is a strong sexual component.

A Bengali legend about the origin of rubies tells of a young man, one of four brothers, whose father died. His mother was passionately fond of him, which made his brothers jealous, so they conspired among themselves to gain control of the estate. Destitute, the youth and his mother set sail one day in a boat.

They sailed down a river and out into the open sea, where eventually they came upon a whirlpool. About this whirlpool were many rubies of enormous size floating on the waves. The youth took one, and then they continued on their journey, finally reaching a distant city where the two settled down to live.

One day the king's daughter saw the ruby and desired it. The king bought it for her, but soon she desired more, and the young man offered to return to the whirlpool to obtain them. In the very same boat he set sail again, and eventually reached his destination.

Determined this time to find the source of the rubies, he left his boat and dove down into the heart of the vortex. At its bottom he came upon a palace standing on the ocean floor. The young man went inside. There he found the god Shiva, eyes closed, lost in meditation, and above Shiva's head a beautiful girl upon a platform. She was sleeping and her head was severed from her body. Out of this wound blood trickled, and when it mixed with the waters of the ocean it turned into rubies which were carried upward by the whirlpool.

Beside the girl were a gold and a silver rod, and as he grasped one her head was magically united with her body. She awoke and warned him of Shiva's anger: "Unhappy young man, depart instantly from this place, for when Shiva finishes his meditations he will turn you to ashes by a single glance of his eye." But he desired her, and they returned together to the world above with many rubies. Ultimately he married both her and the daughter of the king, and they bore him many children.

The sexual element is stronger here. The legend begins with a young man whose father has died and who is passionately loved by his mother: the two live alone to the exclusion of the remainder of the family. It ends with the discovery that the jewels of the vortex are congealed drops of woman's blood. Could they be menstrual blood? Is the whirlpool a vagina?

The appearance of Shiva in this folktale means little to us, but would have triggered in the mind of the Bengali listener a strong set of associations. In Hindu mythology he is portrayed in two contradictory aspects, but both intensely sexual, and often dangerously so. He is the most erotic of gods: "When Shiva married Parvati he made love to her for a hundred celestial years, for he was under the control of his passion, tortured by desire. Seeing this great love-play, the gods were worried that the son born of such a union would destroy the universe." Another myth describes his erotic dance: "The Earth trembled and the tortoise and serpent supporting the Earth could not bear it, but Shiva kept dancing in joy, his eyes whirling. All the gods wondered how he could be made to calm down. . . ."

In his ascetic aspect he is chaste, withdrawn, lost in contemplation —but still erotic. Of his four heads the southern one is chaste and terrible. His penis falls in the forest and is burning hot, for his seed has been accumulating through his meditations for a thousand years. The Sanskrit word *tapas*, used to describe these meditations, means "the heat of asceticism": it was in this highly charged aspect that the young man encountered him at the bottom of the whirlpool.

Psychoanalysts are in the habit of listening for key words, striking and recurrent phrases which enable them to understand their patients. It is also a useful technique in probing a cultural obsession. The place to look is where the general public makes its contact with black holes—in newspapers and newsmagazines. Here is a representative list I have gleaned from a scan of references to them in the literature of the past few years:

ENGULFED IN A BLACK HOLE
PEERING INTO BLACK HOLES
VORACIOUSLY DEVOUR EVERYTHING THEY MEET
BOTTOMLESS PITS
THOSE BIZARRE APERTURES

The message is clear enough.

The Charybdis of Homer's Odyssey opens its maw three times daily to swallow its prey: beside it on a cliff at the ocean's edge grows a fig tree. Edgar Allan Poe's maelstrom snares the unwary fisherman: there, deep in its inky waters, he comes upon the wreckage of still other vessels trapped years ago, hovering poised just above its ultimate throat. A full moon shines down upon the unearthly scene, and a rainbow arches overhead. Our last glimpse of Captain Nemo and his Nautilus is of them battling the whirlpool at the close of Jules Verne's *Twenty Thousand Leagues Under the Sea*. The early maps of Mercator represented the north pole as a mighty vortex into which the oceans of the world steadily poured. The gaping chasm yawns beneath our feet repeatedly in nightmares.

The image of the engulfing void has exercised a fascination over humanity for millennia. Black holes are part of this vast constellation of concerns, and their allure has a strong sexual tinge to it. But sexuality is linked with human creation as well as with engulfment.

The Greek word for hole is a striking one. It is *chaos*, and to the ancient mind the word signified a good deal more than mere turbulence and confusion. It had a strong generative component as well. Ovid refers to chaos in *Metamorphosis* as "a formless void containing the seeds, or potentialities, of all things." Chaos was the original fluid from which all nature flowed.

Out of the wombs of our mothers each of us has emerged, and out of every black hole Hawking radiance floods forth in a pearly light. The black hole is mathematically related to the white hole, a source of endless creation, and the singularity in each is similarly related to the singularity that was the big bang in which all the physical universe had its origin. Each is chaotic, undifferentiated, and flooded with light; a formless void containing the seeds, or potentialities, of all things.

INDEX

Index